AIChE Centennial: ChEmical Engineering Education: Past and Future

Topical Conference at the 2008 AIChE Annual Meeting

Philadelphia, Pennsylvania
16 - 21 November 2008

ISBN: 978-1-61567-203-5

Printed from e-media with permission by:

Curran Associates, Inc.
57 Morehouse Lane
Red Hook, NY 12571

Some format issues inherent in the e-media version may also appear in this print version.

Printed by Curran Associates, Inc. (2009)

For permission requests, please contact AIChE
at the address below.

AIChE
3 Park Avenue
New York, NY 10016-5991

Phone: (203) 702-7660
Fax: (203) 775-5177

www.aiche.org

Additional copies of this publication are available from:

Curran Associates, Inc.
57 Morehouse Lane
Red Hook, NY 12571 USA
Phone: 845-758-0400
Fax: 845-758-2634
Email: curran@proceedings.com
Web: www.proceedings.com

TABLE OF CONTENTS

Two Centuries of Process Safety at DuPont

James A. Klein
E. I. du Pont de Nemours and Company
Wilmington, DE, USA 19898
james.a.klein@usa.dupont.com

DuPont was founded over 200 years ago with a core value for understanding and managing the hazards associated with our processes. From the beginning of manufacturing gunpowder on the Brandywine River, to expansion into chemicals in the early 1900's, to the many, diverse businesses of today, this core value for safety has helped the company reduce risk and prevent serious injuries and incidents. This paper reviews some of this basic corporate history, recognizing that the lessons of the past can help us provide for a safer future.

Introduction

The DuPont Company was founded in 1802 – over two centuries ago – and the foundation of a strong tradition for safety was established that remains a corporate core value today. Thomas Jefferson, the third U.S. President, had encouraged company founder, E. I. du Pont (1772-1834), to manufacture gunpowder since the United States at that time did not have reliable producers of high quality powder [1,2]. E. I. du Pont was almost uniquely qualified, since he had apprenticed with Antoine Lavoisier, the famous French chemist who had also been chief of the French royal powder works. Of course, manufacture of gunpowder was an inherently dangerous business, which E. I. du Pont readily recognized. Reflecting on these dangers, he noted that "we must seek to understand the hazards we live with." While the process hazards of 1802 are very different than the hazards of today, the foundation of the strong current process safety program and performance at DuPont rests in this early corporate history. The continued dedication over 200 years to understanding the hazards present in our processes, and in successfully managing them to prevent serious injuries and incidents, has contributed significantly to current process safety priorities and practices. Learning from experience – what has worked and what hasn't – both in our company, and in the broader chemical industry, is essential for continuous improvement. Therefore, reviewing the sometimes hard lessons of the past can contribute to a safer future.

Foundations of Process Safety on the Brandywine

E. I. du Pont selected a location along the Brandywine River in Delaware for the first gunpowder mills. The river provided water power for the mills, local granite could be used for construction, and the local port in Wilmington could be used for shipping [3]. In his original plans, du Pont noted that "At the side of the stream, there must be sufficient land to distribute the different buildings of the plant at prudent distances from each other [4]." Spacing considerations were key to limiting the effects of possible explosions related to manufacture of gunpowder. Specific instructions included:

> The stamp mill should be at a considerable distance from… any other building. This machine may cause violent explosions and should be surrounded by high and thick walls… It [the graining mill] always contains a somewhat large quantity of finished powder and the necessary presence of the workers may cause most serious accidents, it is therefore important to isolate it as much as possible. [4]

Construction at the Brandywine location was started in 1802, and gunpowder was first sold in 1804 with the name Brandywine Powder (later changed to DuPont Powder) [1,5]. DuPont gunpowder quickly developed a reputation for high quality, and production on the Brandywine continued until 1921.

In addition to careful siting of buildings at minimum safe distances, much consideration was also given to the actual design of the mills. High, thick granite walls were constructed on three sides. The fourth side, which faced the river, and the light gauge roof were designed to direct any explosions towards the river and away from workers and other buildings. Additional design features were intended to prevent powder contamination and potential ignition sources, which could lead to explosions, and to allow workers to exit quickly if needed [6]. Redundant safety valves were also used:

> Two safety valves are used – which are absolutely necessary, as when one only is employed it may by accident get choked and the works explode. [7]

Various barricade designs were also used at other locations as the business expanded.

Members of the du Pont family and supervisors were accountable for the safe operation of the mills, with corporate policy requiring that they be present to startup new equipment or manufacturing processes. The following was posted on the doorway to the mills:

> No employee may enter a new or rebuilt mill until a member of top management has personally operated it. E. I. du Pont. [2]

Management also gave personal instruction on how to safely roll powder kegs and how to avoid dropping them. By 1811, E. I. du Pont issued official safety rules to help ensure safe manufacture:

> As the greatest order is indispensable in the manufacturing as well as for the regularity and the security of works, than the safety of the workmen themselves, the following Rules shall be strictly observed by every one of the men employed in the factory. [8]

Family commitment and accountability were so strong that E. I. du Pont lived at the site. The original plan for the site, in fact, included the requirement that the director's house be constructed "in such a position that the whole plant may be seen from its windows [4]." The du Ponts raised seven children in the house, and on at least one occasion, his wife was injured and the house was severely damaged in an explosion. Ultimately, several family members died of work-related causes [1,2].

Despite precautions, explosions in this dangerous business did occasionally occur. An explosion in 1815 resulted in 9 deaths, which were the first company casualties, and another serious explosion that occurred in 1818 resulted in 34 deaths, which remains the worst process incident in DuPont company history. All accidents were carefully investigated: "the accident cannot be attributed to any stone or nail or other foreign body mixed by chance with the materials [9]" and "This is the only cause which our knowledge of the business enables us to find for the accident [10]." Two years following the major explosion in 1818, changes had been made and it was noted that:

> We have last Thursday another accident in one of our mills which has been attended to with some satisfaction, as nobody was injured and it has proved a fair experiment that upon the plan on which our mills are now built we have not to fear any general explosion like the one which happened here two years ago. From the particular construction of the mill the effects of the explosion have been directed in such a way so as not to communicate to any other part of the works. The whole of our mills are now constructed on that safe plan. [11]

In another instance, it was noted that "3000 lbs. of powder have never before exploded together so innocently and with so little damage [10]." The conscientious efforts of the company to operate safely and to learn from accidents was recognized in 1855 by the Wilmington Morning News:

> The number of persons killed in explosions has been reduced by the persistent efforts of the firm to discover and introduce safeguards to prevent catastrophes. [2]

Lessons from these early years in the company's history are today reflected in many elements of the current process safety program, including safe work practices, training, incident investigation, pre-startup safety reviews, emergency response, and operational discipline.

Formal Safety Organization

In the early 1900's, DuPont was re-organized and aggressively expanded into new chemical businesses, such as dyes, paints, plastics, and cellulose products [1]. In 1911, at a corporate management meeting, it was stressed that "In the manufacture of explosives, we have three important points to consider – safety of operation, quality of product, and the last of all the cost [12]." Later that year, Safety Commissions were formed to study and promote accident prevention measures, first in the High Explosives Operating Department and then in other businesses [12,13]:

> These commissions promulgated formal rules governing, certain operating methods and practices, standard drawings for construction design and mechanical safeguards, injury report forms and methods of reporting and tabulating injuries, plant and equipment inspections, and other miscellaneous safety features. These were all assembled in an indexed code book which was distributed to all plant and operating department heads and other interested persons. [13]

Safety Commission meetings of three operating departments in 1913, for example, required over two weeks and resulted in 300 recommendations.

A corporate Safety Division was established in the Engineering Department in 1915 [12,13] to serve as central clearinghouse for evaluation and purchase of all safety items. The Safety Division acted in an advisory capacity for all plants, which included technical training, safety inspections, and review of all construction projects during the design stage. Prevention of accidents was targeted "by protection, by education, and by eliminating the hazard [12]," but the priority was for eliminating hazards:

> When circumstances will permit, the *elimination* of dangerous conditions is naturally the best course and should receive the first consideration. [14]

About this time, the goal of zero injuries was first mentioned in corporate directives [2].

Over the next 10 to 15 years, the Safety Division was very active making plant inspections, compiling and analyzing accident statistics, conducting special investigations of new processes, and participating in design reviews. In 1926, safety and fire protection were merged to form the Safety and Fire Protection Division. A set of Safety and Fire Protection standards was subsequently issued around 1930, including for example, Safe Practices in Entering Tanks. Detailed, periodic inspections based on these standards were also conducted at plants. The success of these programs is evident from safety statistics collected during this period. Fatality rates were reduced from 5.66 per million hours of exposure in 1910 to 0.09 by 1931. Injury rates and days lost were similarly reduced by 90-95% from 1913 to 1931 [13].

Safety groups were eventually established at all plants. Reflecting early DuPont history, though, responsibility for safety shifted from safety departments at sites to operations management, based on:

> The acceptance throughout the company of the fundamental principle that accident prevention is an integral part of the operating routine and therefore responsibility. [13]

Training programs were offered to all workers, since employee commitment to working safely was recognized as essential for the success of safety programs:

> Recognition of the fact that the safety morale or attitude of the actual workers will mostly govern the success of the safety effort and consequently any program to be successful must be designed to reach every worker through regular organization lines. [13]

One problem, through, was the availability of trained safety engineers and professionals, as expressed by Lewis DeBlois, the first head of the Safety Division and later the head of the National Safety Council, who commented on "a very definite and lamentable scarcity of trained safety engineers technically fitted to carry forward the work [15]." Part of the problem, DeBlois also commented, was that training on safety was not commonly part of most engineering programs:

> … safety engineering, with its interests in design, equipment, organization, supervision, and education… bears as well a very definite and important relation to

> all other branches of engineering. This relation is so close, and its need so urgent, that I am convinced that some instruction in the fundamentals of safety engineering should be given a place in the training of every young engineer. He should be taught to think in terms of safety as he now thinks in terms of efficiency. Conservation of life should surely not be rated below the conservation of energy. Yet, few of our technical schools and universities offer instruction in this subject, and the graduates go out to their profession with only vague surmises on "what all this talk on safety is about. [15]

To help alleviate the scarcity of capable safety engineers "men with the required amount of technical training had to be especially trained for the work, and this was done by and under the supervision of the main office Safety Division [13]." Company safety training remains a key focus today.

Perhaps the most severe challenge to the company's safety focus might be expected to have occurred during the hard economic times of The Depression in the 1930's. However, the company's efforts to understand and control the hazards of its processes continued out of necessity:

> Special studies are continually being made of the hazards arising out of the use of various materials or processes… Where sufficient information is not obtainable… experiments and tests are conducted… [13]

and the core value for safety was re-emphasized:

> The operations of the DuPont Company and subsidiary companies have for many years enjoyed enviable success in safety work; and we believe no other group of plants can point with equal pride to such a high degree of control over a multiplicity of inherent industrial hazards. However, past successes will not prevent present or future accidents. If we are to maintain our position in this field of work, we must not only continue our efforts, but we must increase them. As accidents are reduced, their further reduction becomes increasingly difficult. Furthermore, several factors are present today, undoubtedly due to economic conditions, which make it imperative that our accident prevention work be materially increased particularly in certain directions. [16]

Modern Process Safety Program

In many respects, the modern era of Process Safety Management (PSM) at DuPont began with a serious process incident in Louisville, Kentucky in 1965, which resulted in12 fatalities, 61 injuries, and a property loss of over $50 million. Following the incident, corporate management asked each site to review their production facilities and procedures to assess the potential for catastrophic events and to take appropriate preventive measures. An annual review was also instituted to ensure that process additions or changes did not create new hazards, and in following years, many sites conducted annual, in-depth process hazard reviews to evaluate the process safety of site processes. By 1973, new guidance was issued that detailed suggested hazard review methodologies and frequencies. A corporate guideline for Process Hazards Reviews was issued later in 1978, and a comprehensive, integrated corporate Process Hazards Management (PHM) guideline was issued in 1979.

The PHM guideline, now generally referred to as Process Safety Management (PSM), was intended to help prevent "serious, process-related incidents, which might affect plant personnel, off-site communities, the environment, or result in significant property loss or loss of business." Its purpose was to:

- Establish a framework to help focus management efforts on this important, serious, and complex subject.
- Comprehensively describe the principles and essential features of Process Safety Management for use by sites in managing Process Safety.
- Describe Corporate Business responsibilities and activities.

Business and site management responsibilities were clearly defined, and application of the guideline required that:

> Each site and/or manufacturing facility manufacturing, handling, using, or storing hazardous substances *will*, within the framework of this Guideline, develop a detailed Process Safety Management Program suited to its specific organizational structure and needs.

Several updates to the PSM guideline have occurred since it was first issued, as discussed below.

The incident in Bhopal, India in 1984, involving the release of acutely toxic methyl isocyanate (MIC), resulted in about 2000 deaths and thousands of injuries, significantly raising the awareness of both industry and regulators about the potential for off-site catastrophic incidents [17]. The response in DuPont was immediate. At the time, a product in the agrichemicals business used MIC as a key ingredient in the manufacture of an insecticide. Following Bhopal, the R&D organization quickly developed a new process using inherently safer process principles to allow use of a less hazardous material as the starting raw material, eliminating the risk of shipping and storing MIC. Process upgrades allowed this new raw material to be converted to MIC in the reactor, which was then immediately reacted as part of the next step, limiting the amount of MIC in the process to very low levels [18].

A Highly Toxic Materials (HTM) Subcommittee was also formed to review DuPont operations worldwide. This review found that DuPont's traditional emphasis on management responsibility for managing safety and PSM was effective in minimizing risks both on- and off-site. Additional safeguards were also recommended, though, to provide an extra margin of safety, providing additional guidelines for facilities, storage, operations, transportation, detection, and community protection. In 1985, three new corporate guidelines were issued on Off-Site Risk Assessment, Community Preparedness, and Management of Highly Toxic Materials. The Off-Site Risk Assessment guideline provided "management with information to aid in identifying and assessing potential off-site public exposure," and included risk reduction measures incorporating many principles of inherently safer processes. The Community Preparedness guideline endorsed strong support of the chemical industry Community Awareness Emergency Response (CAER) program and similar efforts in other countries, and it provided guidance on how sites could establish and renew appropriate community organizations to plan and respond to chemical emergencies.

The Management of Highly Toxic Materials guideline identified 15 hazardous materials, such as ammonia and chlorine, where special consideration was required to ensure safe storage, processing, and transportation. Highly Toxic Material (HTM) Safety Guardian Committees were established for managing each hazardous material globally, and specific principles and guidance were provided on facilities, storage, operations, transportation, detection, and community protection. For example, one principle for detection was:

> Sites handling highly toxic materials will include use of automatic or continuous specific chemical detection systems, where appropriate, as an integral part of minimizing the potential for harmful exposure on- and off-site.

Each HTM Safety Guardian Committee used this guidance to develop specific standards and practices to help ensure safe use of each material. A HTM Safety Guardian Leadership Team was formed in 1988 to "provide continued global focus, leadership, coordination, and leveraged support for the safe management of Highly Toxic Materials." Upgrades to HTM guidance over the last 20 years have included audit requirements, product stewardship reviews, and additional guidance on minimizing storage and preventing, detecting, controlling, and containing releases. HTM Safety Guardian Committees now exist for over 20 substances.

In addition to corporate PSM and HTM programs, individual businesses developed specialized process safety programs, often related to reactive chemicals, to help ensure the safety of site operations. In many cases, these programs resulted from serious process incidents, leading to formation of global teams from multiple manufacturing sites with similar processes to help prevent further incidents [19]. These teams shared technical and process safety information, developed specific guidance and standards for promoting safe operations, developed hazard evaluation methodologies, provided training, and developed audit protocols. In some cases, these programs pre-dated the issuance of corporate PSM guidance (e.g., one program began in 1968 and another in 1975) and are still active today, continuing to contribute to the prevention of serious process incidents.

In late 1996, the corporate ZIP (Zero Incidents… Period!) Team was formed to focus on eliminating process-related injuries and incidents, while at the same time, building business value for PSM. The ZIP Team reaffirmed the vision of "The Goal is Zero" for process incidents and communicated these key messages:

- There is no silver bullet, no single answer, to improving process safety.
- The goal has to be ZERO process-related injuries and incidents. The question is, not IF, but HOW, this goal can be achieved, just as it is for personal safety.
- PSM is a business issue, not only a manufacturing issue, requiring the contributions of a broad cross section of the organization to ensure success.
- Businesses therefore need to establish a sustainable continuous improvement process toward achieving the goal of zero PSM injuries and incidents that is embedded in business planning.

All businesses and regions were requested to have implementation plans by the end of 1997, which would be included in executive level discussions on performance and planning. The result was a renewed corporate effort to drive continuous improvement activities that led to upgraded standards and practices, training programs, evaluation tools, and business metrics that have greatly impacted

PSM performance in DuPont. Subsequent improvement activities, including a PSM Discovery Team effort in 2006, have helped to improve PSM programs on a continuous basis.

Looking Forward

In thinking about the future, the one certainty is that changes will occur. Businesses will grow, acquisitions will be made, and new technologies will be invented. Process safety will need to change as well. New acquisitions may present challenges due to different corporate or geographic safety cultures. New technologies may benefit from the application of PSM, but at the same time, may require that new approaches be developed. Standing still and celebrating our successes, though, will not work. Continuous improvement will always be necessary:

> Systems and organizations continually experience change as adaptations are made in response to local pressures and short-term productivity and cost goals… A corollary of this propensity for systems and people to adapt over time is that safety defenses are likely to degenerate systematically through time, particularly when pressure toward cost-effectiveness and increased productivity is the dominant element in decision making. Thus, the redundancy and other precautions added to protect against human error often degenerate over time as other work practices adapt to increase efficiency within the local environment. The critical factor here is that such adaptation is not a random process – it is an optimization process… [20]

The challenge is to recognize this problem and continually work to maintain robust process safety programs. These words, written by a DuPont safety manager over 75 years ago, remain true today:

> However, past successes will not prevent present or future accidents. If we are to maintain our position in this field of work, we must not only continue our efforts, but we must increase them. [16]

The goal then, now, and in the future is to prevent serious injuries and catastrophic incidents – the goal is zero. Learning from experience, based on the lessons of the past, is essential if zero is to be achieved. At DuPont, our core value for safety was established over 200 years ago by the founder, E. I. du Pont, and it continues today. Seeking to understand and manage the hazards of our processes to help ensure a safe future remains our challenge and our commitment.

References

1 Kinnane, Adrian, *DuPont: From the Banks of the Brandywine to Miracles of Science*, E. I. Du Pont de Nemours and Company, 2002

2 Mottel, William J., Joseph F. Long, and David E. Morrison, *Industrial Safety is Good Business: The DuPont Story*, Van Nostrand Reinhold, 1995

3 Carisio, Justin, *History of DuPont*, DuPont video, 2003

4 du Pont, E. I, *The Location and Construction Necessary for Manufacturing Gunpowder*, Hagley Museum and Library, Wilmington, DE, ca 1801

5 Johnson, Allen and Dennis Molene, *Dictionary of American Biography, Vol. 5*, Charles Scribner's Sons, Hagley Museum and Library, Wilmington, DE, 1930

6 Cassell's Magazine, *In a Powder Mill,* Hagley Museum and Library, Wilmington, DE, ca 1890

7 Du Pont, B. G, *Life of E. I. du Pont, Vol. 6, 1802-04*, University of Delaware Press, Hagley Museum and Library, Wilmington, DE, 1925

8 du Pont, E. I., opening paragraph to "Rules" for his manufactory, Hagley Museum and Library, Wilmington, DE, 1811

9 DuPont, *Letter Books, 1820-21*, Hagley Museum and Library, Wilmington, DE

10 DuPont, *Letter Books, 1825-27*, Hagley Museum and Library, Wilmington, DE

11 du Pont, E. I, letter to a customer in Virginia, Hagley Museum and Library, Wilmington, DE, 1820

12 Petersen, Peter B., *Lewis A. DeBlois and the Inception of Modern Safety Management at DuPont, 1907-1926*, submitted to Academy of Management, Management History, Division, Hagley Museum and Library, Wilmington, DE, ca 1987

13 DuPont, *Development of Accident Prevention Work in E. I Du Pont De Nemours and Co.*, Hagley Museum and Library, Wilmington, DE, 1932

14 DeBlois, L. A., *The Application of Safety Devices*, Hagley Museum and Library, Wilmington, DE , 1916

15 DeBlois, L. A., *The Safety Engineer*, American Society of Mechanical Engineers, Hagley Museum and Library, Wilmington, DE, 1918

16 Miner, H. L., *Accidents Are Increasing*, letter, Hagley Museum and Library, Wilmington, DE, 1933

17 Atherton, John and Frederic Gil, *Incidents That Define Process Safety*, Center for Chemical Process Safety, John Wiley & Sons, 2008

18 Clements, C. Curtis, *Application of Inherently Safer Process Concepts in DuPont*, Paper TH002a, Plant Process Safety Symposium, Atlanta, April 11-13, 2005

19 Balchan, A. S., J. A. Klein, and F. G. Klein, *Process Safety of Polymer Resin Manufacturing: A 20-Year Perspective*, in Loss Prevention and Safety Promotion in the Process Industries, Vol. 1, Elsevier Science, 1995

20 Leveson, Nancy, *A New Accident Model for Engineering Safer Systems*, Safety Science, Vol. 42, p. 237-270, 2004

8b The Process Safety Journey at Air Products

Shakeel Kadri, Air Products & Chemicals Inc., 7201 Hamilton Blvd, Allentown, PA 18104

We at Air Products have developed a very robust "safety" culture over the years. My talk will review the highlights of the journey that we started in the mid seventies following two very tragic process safety incidents. It will discuss the approach our senior leadership then took, that has become one of the cornerstones of Air Products safety philosophy, which is designed to keep our own people and others affected by our operations safe. The talk will also review the lessons we learned, our current focus on how we manage and execute our process safety requirements on a global basis, and discuss the unique challenges and opportunities we face in the new world.

8c Process Safety at OSHA

Lisa Long, Occupational Safety and Health Administration, 200 Constitution Ave, NW, Room N3119, Washington, DC 20210

This presentation will provide a brief history of OSHA's efforts in process safety management. It will include discussion of the PSM standard and other compliance efforts including the combustible dust and petroleum refinery NEP's.

8d History of Process Safety and Loss Prevention In the American Institute of Chemical Engineers

Dennis C. Hendershot, Chilworth Technology Inc., 930 Greenhouse Dr, Bethlehem, PA 18017

Process safety has always been an important consideration in chemical engineering, and the activities of the American Institute of Chemical Engineers (AIChE) have expanded over the years to more formally focus on this critical aspect of chemical engineering practice. Early initiatives included the annual symposia on Safety in Ammonia and Related Plants, dating back to the early 1950s, and the Loss Prevention Symposium, first held in 1967. These activities led to the formation of the AIChE's Safety and Health Division in 1969, and the division assumed responsibility for these existing programming activities. Following the Bhopal tragedy in India in 1985, AIChE formed the Center for Chemical Process Safety (CCPS) as an industry alliance to share and enhance process safety expertise in the industry. One of the early CCPS activities was the creation of the Safety and Chemical Engineering Education (SACHE) program to enhance teaching of process safety in the university chemical engineering curriculum. This paper will review the history of these ongoing activities, including many observations from interviews and contributions from people who help to organize them.

8e Preventing Accidents through Experience Based Learning

Scott Berger, American Institute of Chemical Engineers, New York, NY 10016

On December 3, 1984, water contamination of a tank of methyl isocyanate in Bhopal, India initiated a series of events that led to a catastrophic toxic release, killing more than 3,000 residents and injuring over 100,000. Immediately after, leaders from the chemical industry asked AIChE to lead a collaborative effort to eliminate catastrophic process incidents by advancing state of the art technology and management practices, serving as the premier resource for information on process safety, supporting process safety in engineering, and promoting process safety as a key industry value. In the spring of 1985, CCPS was founded. Since that time, CCPS has incorporated experience-based learning from major and minor accidents and near-misses into a world-class collection of guideline documents, resource books, and on-line content. This paper will reflect on the history of CCPS, highlight some recent accomplishments, and present a glimpse at the future of process safety.

Safety and Chemical Engineering Education – History and Results

(J. F. Louvar, CCPS Staff Consultant for SACHE)

Extended Abstract

CCPS started their Safety and Chemical Engineering Education (SACHE) efforts in 1985. The objective of CCPS's SACHE is to help professors add safety to their chemical engineering courses. This added education is needed because large chemical plant accidents continue to increase every year. The consequences are deaths, injuries, and environmental damage, and there is currently no sign that there will be fewer accidents in the future. The causes are the complexity of chemical processes, the aging of our existing plants, and sizes of your newer plants. Our students and engineers must be motivated to prevent these accidents in this complex environment.

The ultimate goal of SACHE (CCPS and Industry) is to create a "safety culture" in our young professionals by adding elements of safety to the undergraduate curriculum. The idea is that if students are exposed to safety in small quantities throughout their education, then it will become a part of their inherent thinking both as students and when working as professionals. This safety culture will help our future professionals be motivated to design, install, and manage safe plants in this challenging environment. Clearly the detailed education on specific process safety topics is the responsibility of industry; however, the effectiveness of this training will be substantially improved when working with a positive safety culture.

This SACHE committee's twenty-three years of hard work is now giving a significant number of students and young engineers an interest and knowledge in chemical process safety. CCPS/SACHE is making progress as shown below:

- **Motivating professors regarding the importance of process safety (annual workshops for professors in a plant environment).** For these workshops, professors only pay for the transportation to and from the host plant. Local transportation, food, lodging, etc. are funded by CCPS and Industry. The workshops are facilitated by experts in the area of chemical process safety. Plant tours are a major part of this experience to highlight the teachings. This is a positive and effective educational experience for the professors (25 per annual workshop), and a unique opportunity to tour plants with plant specialists.

- **Helping professors add safety to their courses (SACHE develops course materials).** The SACHE products are on-line, making them readily accessible to professors and students. The products include movies and presentations on: a) chemical process safety, b) inherent safety, c) risk assessment, d) static electricity, e) case histories, etc.
- **Motivating students regarding the importance of process safety (AICHE Design Problem Awards and SACHE Safety Certificates).** There are six awards for the AICHE Design Problem Solutions. The awards are SACHE awards ($200 for an individual and $300 for a team that includes the best application of the principles of process safety in their design solutions), and the Safety and Health Division Awards (four at $600 each for the best application of the concepts of inherent safety in their design solutions). SACHE also has safety certificates covering the major topics of process safety. Students are directed to review specific SACHE products and then are required to take a test. Students who pass the test are given certificates – and the certificate awardees are acknowledged via AICHE. The awardees can and should identify the certificates in their resumes, because these awards are and will continue to be especially valued by industry and interviewers.

62a Molecular Concepts In Thermodynamics Education

David A. Kofke, Chemical and Biological Engineering, University of Buffalo, 303 Furnas Hall, SUNY at Buffalo, Buffalo, NY 14260

The laws of thermodynamics emerge from the molecular basis of matter, but one of the strengths of thermodynamics is that it needn't acknowledge the existence of molecules to quantify the behavior that results from molecular interactions. Nevertheless it can be difficult to gain an intuitive understanding of thermodynamics, and to make predictions about new systems, without a molecular understanding. Molecular concepts are becoming all the more relevant the with growth of nano- and bio-technology as key elements of chemical engineering. Thus there can be significant benefits to including molecular concepts in chemical thermodynamics education. The challenge is to accomplish this without greatly diluting the treatment of traditional and proven thermodynamics methods. This presentation will describe past, present, and future roles of molecular concepts in thermodynamics education.

62b Concepts in a Chemical Engineering Thermodynamics Course

John L. Falconer, University fo Colorado, 1111 Engineering Drive, Boulder, CO 80309-0424

Many concepts that are important to chemical engineering are introduced and used in thermodynamics courses, but students often solve quantitative problems without having a functional understanding of the concepts involved. Instead, they memorize an algorithm for solving such problems. Qualitative questions that address the important thermodynamic concepts can be used in class, on assignments, and on exams to challenge students with questions that cannot be answered by memorization. These in-class questions are at a level so that only approximately one-third of the students answer correctly initially, even after reading the appropriate sections of the book and discussing the concept in class. Good problems require good wrong answers that show the misconceptions many students have about the material. Examples will be presented to demonstrate the lack of understanding of some of the important concepts in chemical engineering thermodynamics, such as vapor pressure, phase equilibrium, phase changes, reversible and irreversible processes, entropy, fugacity, state functions, chemical equilibrium, ideal gases, energy balances, and cycles.

Many of these conceptual problems use diagrams that represent various behaviors, and many wrong answers are obtained from student answers to conceptual exam questions. These questions are used effectively in class to replace much of the lecture. They are used in combination with student-held remote RF transmitters (clickers) and peer instruction. Students overwhelming like this mode of instruction.

62c Visualization: Exposing the Art in Thermodynamics

Kenneth Jolls, Chemcial Engineering, Iowa State University, 2155 Sweeney Hall, Ames, IA 50011

The visual element in classical thermodynamics was affirmed 135 years ago when J. Willard Gibbs used imagery to show how the criteria for equilibrium were revealed in a model of the surface of the Fundamental Equation. The torch that Gibbs lit has been carried by many scholars of the subject since then who have used a variety of ways to draw pictures of thermodynamic subjects. We are fortunate today to have the computer to manage the details of graphics creation so that we—the teachers of thermodynamics—can manage the content. The possibilities are enormous. "The "Medium can Bring the Message."

This talk will present both old and new examples of thermodynamic information visualized and touch also upon the use of graphics in other areas of chemical engineering education. Half a century ago Edward R. Murrow predicted that "Television can teach." But some would argue today that his prophecy has only partially come to pass. The impressive graphics tools of the 21st century suggest the same for computer visualization. How the future will treat that notion lies largely in the hands of educators like those who will attend this talk.

A Survey of Thermodynamics and Transport Properties in Chemical Engineering Education in Europe and the USA

Peter Ahlström[1,*], Karel Aim[2], Ralf Dohrn[3], J.Richard Elliott[4], George Jackson[5], Jean-Noel Jaubert[6], Maria Eugénia Rebello de A. Macedo[7], Juha-Pekka Pokki[8], Kati Reczey[9], Alexey Victorov[10], Ljudmila Fele Žilnik[11] and Ioannis Economou[12]

1 University of Borås, SE-50190 Borås, Sweden; Email: Peter.ahlstrom@hb.se
[2]Academy of Sciences of the Czech Republic, Prague, Czech Republic, [3]Bayer Technology Services GmbH, Leverkusen, Germany, [4]University of Akron, Ohio, USA, [5]Imperial College, London, UK, [6]University of Nancy, France, [7]University of Porto, Portugal, [8]Helsinki University of Technology, Finland, [9] Budapest University of Technology and Economics, Hungary, [10]St Petersburg State University, Russia [11]National Institute of Chemistry, Ljubljana, Slovenia, [12]National Center for Center for Scientific Research 'Demokritos', Institute of Physical Chemistry, GR-153 10 Aghia Paraskevi, Attiki, Greece
*Presenting author

Introduction

Thermodynamics and Transport Properties (TTP) is a central subject in the majority of chemical engineering curricula worldwide and it is thus of interest to examine how TTP is taught today in various countries. The contents and the organization of the courses reflect an understanding of reality (or philosophy) which often is not explicitly expressed. The discussion about different learning styles (cf., e.g., [1]) and their implication on teaching methods has also spurred us to investigate which methods are used for thermodynamics teaching, especially as thermodynamics often is regarded as a "difficult subject". Our ultimate aim is that our results will help to improve chemical engineering education.

A survey of graduate thermodynamics education in the USA only was performed a few years ago by Visco et al. [2]. As far as we know no systematic study of the undergraduate thermodynamics education has been performed, at least in recent years. In the present study, a survey about TTP education in both undergraduate and graduate programs in Europe and the USA is presented. Answers from 136 universities from twenty different European countries and the USA were collected and used for this study. This study was performed under the auspices of the Working Party of Thermodynamics and Transport Properties of the European Federation of Chemical Engineering.

Methods

The survey was performed using a web based surveying system (MrInterview of the SPSS package) for which invitations were sent out to the universities. Several reminders were sent out but the answer frequency was varying significantly between the different countries (cf. Table 1).

Table 1. Number of universities/colleges per country that responded to the survey. No answers were obtained from any other country*).

Country	No of answers	Country	No of answers
Austria	1	Italy	5
Belgium	3	Netherlands	1
Bosnia and Herzegovina	1	Norway	1
		Portugal	5
Croatia	2	Russia	2
Denmark	2	Serbia	1
Estonia	1	Slovenia	2
Finland	4	Sweden	4
France	6	United Kingdom	7
Germany	28	USA	55
Greece	3		
Hungary	2		
		TOTAL	136

*)Countries from which no university / college responded are not shown.

Results

Of the universities that answered more than 70 % offer B.Sc. education, 65 % offer M.Sc. education and 55 % offer Ph.D. education. Most universities offer at least two courses of thermodynamics. Half of these are taught to chemical engineers exclusively whereas the rest are taught with other branches of engineering, mainly mechanical and / or process engineering.

Thermodynamics teaching in terms of quantity

The amount of thermodynamics taught has been analyzed both with respect to the number of courses and their size. The number of courses reported from each university is given in Table 2 where it can be seen that a majority of the universities in Europe report more than two courses each whereas in the USA the majority reports at most two courses. Hence, most of the following discussion will be based on the first two courses reported from each university.

Table 2 Number of thermodynamics courses reported at various universities

No of courses	% Europe	% USA
1	21	35
2	25	41
3	22	14
4	14	7
5	9	2
> 5	10	0

An issue that caused much confusion among the respondents was the definition of course size since no unambiguous measure of course length and work load exists. We have chosen to use the work load measured by the amount of full time study weeks per course, i.e., the intention is that if a course was expected to be studied as the only course during a given period, the value given should be the number of weeks that course was expected to fill the students time. If the student in a given program was expected to follow two courses at the same time, the week should be split between the courses according to their generated work load. In order to simplify the calculation for European universities we introduced a transformation from the European Credit Transfer System (ECTS) introduced with the Bologna agreement in Europe. 1.5 ECTS units equals approximately one week of work since one year usually contains approximately 40 study weeks and should correspond to 60 ECTS units. However, judging from the reactions of the respondents, such a measure is not familiar in many countries yet, and thus some care has to be exercised when interpreting the results regarding course lengths.

Table 3. Amount of thermodynamics (as full time study weeks) forming part of different course(s)

	Chem.Eng		PhysChem		Independ.		Other	
N:o of weeks	% USA	% Eur	% USA	% Eur	% USA	% Eur	% USA	% Eur
None	-	8	18	31	56	31	22	28
<3	2	8	37	34	4	20	41	21
3-7	43	26	32	23	24	20	24	32
8-12	9	32	7	6	4	15	11	14
13-18	13	15	4	6	4	8	2	5
19-24	9	5	2	1	4	4	-	1
>24	24	11	-	-	6	2	-	-

"Chem.Eng." does not include physics, physical chemistry and similar fundamental courses but only (thermodynamics in) the applied chemical engineering courses.

In both regions, somewhat more than 40% spend at most 7 weeks on thermodynamics. In Europe the courses are generally less than 19 weeks whereas in the USA, one fifth spend more than one semester on thermodynamics. In general two sets of course lengths were observed, corresponding either to a full semester of full time studies or to quarter of a semester.

Table 4. Contents of thermodynamics course 1 (percent of total number of responses)

Topic	Central		Treated in some detail		Mentioned		Not part of the course	
	Eur	USA	Eur	USA	Eur	USA	Eur	USA
1st law	90	91	8	7	0	2	2	0
2nd law	88	80	10	11	1	2	1	7
Entropy	80	74	14	13	4	6	3	7
Molecular/Statistical interpretation of entropy	9	9	24	15	35	48	32	28
Free energy and quality of energy	44	43	22	26	22	19	11	13
3rd law and absolute entropy	26	33	21	18	35	35	18	33
Thermodynamic cycles	55	50	28	37	8	7	10	6
Heat expansion of solids and liquids	14	18	34	30	33	35	20	17
Equations of state	45	56	36	32	10	11	9	2
Phase equilibria	39	48	26	15	16	9	19	28
VLE	30	46	21	18	21	4	28	32
LLE	15	22	18	19	15	19	52	41
Heat transfer	9	7	19	11	20	39	52	43
Thermochemistry	21	9	16	20	6	30	56	41
Statistical thermodynamics	5	2	4	6	26	30	65	63
Molecular simulation	1	0	1	7	14	15	84	78
Kinetic theory of gases	8	0	15	9	32	22	45	68
Non-equilibrium thermodynamics	3	2	12	2	12	9	78	87
Thermodynamics for biological systems	4	0	3	6	16	37	78	58

Table 5. Contents of thermodynamics course 2 (percent of responses for course 2)

Topic	Central		Treated in some detail		Mentioned		Not part of the course	
	Eur	USA	Eur	USA	Eur	USA	Eur	USA
1st law	33	43	17	17	27	20	23	20
2nd law	36	46	14	20	25	17	25	17
Entropy	28	49	22	23	23	14	27	14
Molecular/Statistical interpretation of entropy	11	11	16	26	34	34	39	29
Free energy and quality of energy	36	34	19	37	22	14	23	14
3rd law and absolute entropy	19	11	25	23	22	29	34	37
Thermodynamic cycles	34	6	5	17	13	34	48	43
Heat expansion of solids and liquids	11	11	27	17	19	43	44	29
Equations of state	56	51	16	40	8	6	20	3
Phase equilibria	59	78	11	11	13	9	17	3
VLE	52	78	14	11	8	6	27	6
LLE	42	54	12	17	8	20	38	9
Heat transfer	22	6	12	17	14	20	52	56
Thermochemistry	36	29	17	31	6	23	41	17
Statistical thermodynamics	8	14	9	14	16	40	67	31
Molecular simulation	3	3	3	9	17	46	77	43
Kinetic theory of gases	8	3	8	14	23	26	61	57
Non-equilibrium thermodynamics	3	0	9	9	11	14	77	77
Thermodynamics for biological systems	3	3	8	17	6	31	83	49

In the first course, the first and second laws of thermodynamics as well as entropy are central in both regions (but 7% in the USA do not mention entropy!). Normally the statistical interpretation of entropy is mentioned as well as the third law and absolute entropy but not in significant depth. The second course is frequently more concentrated on phase equilibria. Both of these courses mainly consist of classical thermodynamics whereas the molecular interpretation often is touched upon. Statistical thermodynamics and molecular simulation as well as thermodynamics for biological systems are not central in any of the two courses neither in USA nor in Europe, but they are more frequently mentioned in the USA. Non-equilibrium themodynamics is equaly not part in any of the two first courses in any of the regions. These results are also reflected in the choice of course books, cf Table 6.

Table 6. The most popular textbooks for course 1 and 2
Books by the same (team of) authors have not been separated since the exact version often is unclear from the answers. There is a obvious difference in the choice of course books between the two continents even though a few books are popular in general, e.g., Smith, van Ness & Abbott.

Course 1	Europe	USA
Atkins (several versions)	18%	
Smith, van Ness & Abbott	11%	39%
Baehr et al.	8%	
Cengel (several versions)	4%	2%
Elliott & Lira		14%
Sandler		13%
Felder		11%
Koretsky		7%
Course 2		
Atkins (several versions)	23%	3%
Prausnitz (several versions)	11%	
Baehr	8%	
Smith, van Ness & Abbott	8%	43%
Gmehling	6%	
Sandler		14%
Koretsky		6%

Table 7. Time (in hours/course)used for different forms of teaching in course 1 in Europe [percent of answers; "(outside)" means "expected student work outside class", PBL = Problem Based Learning, cf. [1].]

Type	0 h	1-20 h	21-40 h	41-60 h	>60 h
Lectures (in class)	-	16	48	25	11
Lectures (outside)	16	39	28	11	5
Exercise classes	8	44	40	6	2
Exercise class (outside)	20	36	32	9	1
PBL etc. (in class)	70	25	5	-	-
PBL etc. (outside)	66	29	4	1	-
Home assign (in class)	41	44	8	6	1
Home assign (outside)	34	35	16	1	3
Laboratory classes	66	25	4	3	2
Lab classes (outside)	78	18	3	1	1

Table 8. Time used for different forms of teaching in course 1 in the USA (in hours/course), [Percent of answers, cf Table 7]

Type	0 h	1-20 h	21-40 h	41-60 h	>60 h
Lectures (in class)	-	11	54	30	6
Lectures (outside)	43	33	11	7	6
Exercise classes	11	59	22	7	-
Exercise class (outside)	48	35	13	2	2
PBL etc (in class)	26	56	13	4	2
PBL etc (outside)	44	32	15	7	2
Home assign (in class)	20	37	28	9	6
Home assign (outside)	11	28	20	20	20
Laboratory classes	83	15	-	2	-
Lab classes (outside)	83	13	4	-	-

Table 9. Time used for different forms of teaching in course 2 in Europe (in hours/course) [Percent of answers, cf Table 7]

Type	0 h	1-20 h	21-40 h	41-60 h	>60 h
Lectures (in class)	3	19	46	22	10
Lectures (outside)	18	35	25	3	5
Exercise classes	10	38	38	10	5
Exercise class (outside)	24	32	27	13	5
PBL etc (in class)	67	27	-	5	2
PBL etc (outside)	73	22	5	-	-
Home assign (in class)	48	40	8	3	2
Home assign (outside)	38	32	18	2	11
Laboratory classes	73	16	6	2	5
Lab classes (outside)	81	16	3	-	-

Table 10. Time used for different forms of teaching in course 2 in the USA (in hours/course) [Percent of answers, cf Table 7]

Type	0 h	1-20 h	21-40 h	41-60 h	>60 h
Lectures (in class)	-	16	48	25	11
Lectures (outside)	16	39	28	11	5
Exercise classes	8	44	40	6	2
Exercise class (outside)	20	36	32	9	1
PBL etc. (in class)	70	25	5	-	-
PBL etc. (outside)	66	29	4	1	-
Home assign (in class)	41	44	8	6	1
Home assign (outside)	34	35	16	1	3
Laboratory classes	66	25	4	3	2
Lab classes (outside)	78	18	3	1	1

Normally courses are centered around lectures and exercise classes with little or no laboratory work whereas home assignments are given in the vast majority of the courses. It can be noted that a rather large amount of time is used in class for home assignments and that problem based learning is twice as popular in course in the USA as in Europe.

Conclusions

In general, the fact that classical thermodynamics has not changed profoundly during the past decades is reflected in the invariability of the thermodynamics courses. The most popular textbook had its first edition 60 years ago and most other textbooks follow the same outline. More modern atomistic viewpoints are normally found in the (elective) later courses where they often are combined with statistical thermodynamics. This result may reflect how thermodynamics is needed and used in industry and an investigation of the use of thermodynamics within industry is also on-going within the Working Party. The results of that will be reported in the near future as well as a more detailed analysis of the education.

Since the response frequency is not extremely high, caution is needed when drawing conclusions from the material but we assume that some results are clear. Even though the results are quite similar for the USA and Europe a notable difference is the higher amount of problem based learning as well as the larger work with home assignments in the USA. Also there seems to be some more emphasis on atomistic understanding in American teaching.

Acknowledgements

Support by Peter Sigrén at CLU University of Borås, with the SPSS interview system “MrInterview” is gratefully acknowledged.

References

[1] P.Ramsden, Learning to teach in higher education, 2nd ed. London : Routledge, 2003.
[2] S.K.Dube, D.P. Visco, Chem. Eng. Ed., 2005, 258-263.

62e Biological Applications In Thermodynamics Education

Stanley Sandler, University of Delaware, Department of Chemical Engineering, Newark, DE 19716

The field of chemical engineering has evolved into areas of biochemical engineering, nanotechnolgy, product design, and other emerging fields. Applied thermodynamics plays a surprisingly important role in all of these areas. This presentation will focus on some of the applications of thermodynamics to biological chemical applications, and how these can be brought into the classroom.

WAS A TRANSPORT PHENOMENA COURSE IN CHEMICAL ENGINEERING CURRICULA INEVITABLE?

R. Byron Bird
Chemical and Biological Engineering Department
University of Wisconsin-Madison
Madison, WI 53706 U.S.A.

This account covers primarily the years 1941-1961, which is by now "ancient history" in chemical engineering. In 1941, the United States became involved in the Second World War, and I enrolled as a freshman in ChE at the University of Maryland.

§1. Teaching of transport phenomena before and after WWII

During the period 1941 to 1951, the subject of transport phenomena was not taught to undergraduates as a unified subject in any U.S. university. The *flow of fluids* was covered at an elementary level in a sophomore physics course; this included the Bernoulli equation, the definition of viscosity, Poiseuille's law, qualitative ideas about turbulence, and other assorted topics. The *flow of heat by conduction and by radiation* was also taught in sophomore physics; but usually free and forced convection were taught only qualitatively, and heat production by viscous dissipation and chemical reactions was not included at all. The *diffusion of chemical species* was generally covered in physical chemistry, but the mass flux and the diffusion coefficient were poorly defined; convection, thermal diffusion, and forced diffusion were not taught at all. After 1945, physics departments were focusing their attention on topics in nuclear physics and losing interest in the "classical" physics topics; similarly chemistry departments were moving into new areas, and diffusion was of marginal interest.

The two unit operations books used at the time were the third edition of *Principles of Chemical Engineering*, by Walker, Lewis, McAdams, and Gilliland (1937)[26] and the second edition of *Elements of Chemical Engineering*, by Badger and McCabe (1936)[1]. These books were applications oriented and made little use of the principles of transport phenomena. Students trained in this era were able to solve important practical problems by following examples that were in the textbook, but were unable to undertake studies outside of that scope.

Olaf Hougen realized that the subject of transport phenomena was of interest in connection with the future development of many topics in chemical engineering. In the spring of 1950, he persuaded Professors J. O. Hirschfelder and C. F. Curtiss of the Chemistry Department to teach a graduate course in transport phenomena in the Chemical Engineering Department. This course covered roughly the material that appeared later in Chapter 11 of *Molecular Theory of Gases and Liquids*, by Hirschfelder, Curtiss, and Bird (1954)[16]. This course was never taught again.

§2. Early developments

Transport phenomena, as a discipline, definitely bears a "made in Europe" label. The mathematical development of the kinetic theory of gases by Maxwell (in England) and Boltzmann (in Austria) in the 19th century laid the cornerstone for the systematic discussion of mass, momentum, and energy transport; this was summarized in the classic book of Chapman and Cowling (in England), titled *The Mathematical Theory of Non-Uniform Gases* (1939)[13]. There it was shown how to derive the equations of change for mass, momentum, and energy from molecular arguments, and to derive expressions for the fluxes and transport properties in terms of the intermolecular forces.

The continuum theory of transport phenomena began much earlier with the flux of momentum (stated by Newton in England in the 17th century), the flux of energy (stated by Fourier in France early in the 19th century), and the flux of mass (by Fick in Germany in the middle of the 19th century); the radiant energy flux was given by Planck in Germany (at the end of the 19th century).

The equations of change were obtained by continuum arguments by a rather slow process, but these equations were complete by the beginning of the 20th century. An early summary of the "field equations" of transport phenomena was given by G. Jaumann (a Rumanian by birth), who in a review article in *Sitzungsberichte Akad. Wiss. Wien* (1911)[17], presented many of the basic equations found in transport phenomena textbooks today. It is unfortunate that his name is all but forgotten, and that biographical material on him is very limited (the only biographical information I know about is an article by H. Bednarczyk[2] in 1990).

The German literature on transport phenomena ("Transporterscheinungen," in German) and engineering applications was quite well developed. The extensive research and authoritative books by such giants as Prandtl, Schlichting, Nusselt, Stefan and others attest to the tremendous scholarship that was produced in Germany in the era before and between the two world wars.

An early book in English covering roughly the material that chemical engineers would find useful was R. C. L. Bosworth's *Transport Processes in Applied Chemistry: The Flow of Physical Properties in Chemical Reactors*, Pitman (1956)[12]. However, it could not be used as a textbook, since there were no illustrative examples and no unworked problems. In about 400 pages, he discusses a very wide range of topics. Bosworth's book resembles more a long review article, complete with references; however, no topic was treated in depth (e.g., the Navier-Stokes equation of motion of fluid dynamics is given in vector form in a short discussion of less than one page).

§3. Rumblings and fumblings at the University of Wisconsin

This section covers events in Madison from 1953 to 1957. The year 1953 marks the arrival of ENL and RBB in the Chemical Engineering Department at the University of Wisconsin. ENL had gotten a five-year BChE and a PhD from Cornell University and then worked for the Charles Pfizer Company in Brooklyn, NY; RBB had received a BS in ChE from the University of Illinois, and a PhD in Chemistry at the University of Wisconsin, followed by a postdoctoral year in theoretical physics at the University of Amsterdam, a year of teaching in the Chemistry Department at Cornell University, followed by a summer at the DuPont Experimental Station. (WES joined the department later, in 1956, after a BS and an MS at the University of Wisconsin and a ScD at MIT; then he worked at Sinclair Research Laboratories for about five years.)

Bob Marshall had been planning to teach a fluid dynamics course in the fall of 1953 using NACA reports by H. Schlichting and others. When Bob was promoted to Associate Dean of the college, I was asked to take over that course. I ended up using Schlichting's *Grenzschicht-theorie* (*Boundary Layer Theory*)[22] as the textbook for the course. By doing that, I became familiar with the rich experimental and theoretical literature on fluid dynamics in German. I had already had some contact with this literature at DuPont where I had been asked to study the heat production by viscous heating in polymer extrusion.[3a] None of the chemical engineers that I came into contact with at DuPont could explain where the viscous heating term in the energy equation came from. It was apparent that that chemical engineers at that time were not getting enough fundamental training in fluid dynamics to tackle problems outside of the problems that were in the unit operations textbooks.

While teaching the fluid mechanics course, I got interested in the "mechanical energy balance," which I could not understand from the unit operations textbooks. It seemed to me that all the macroscopic balances should be obtained by integrating the equations of change. Indeed, I found that the mechanical energy balance could be obtained as an integral over the flow system of the dot product of the local velocity with the equation of motion. This led to the publication in *Chemical Engineering Science* of a paper on the derivations of the macroscopic balances.[4]

The following term I taught the graduate course titled "Diffusional Operations," using Sherwood and Pigford[23], *Absorption and Extraction*. This was quite instructive, since this material was all new to me. By the end of the semester, I had worked through a lot of the basic background material needed for the teaching of the applications. Since the scientific background material was not readily available anywhere, I prepared a summary article and sent it to Academic Press, for publication in Vol. 1 of *Advances in Chemical Engineering* (1956)[5]. This article established the notation that would be used later in "BSL" and presented a number of illustrative problems as well; great care was taken with respect to the definition of diffusivity, the frames of reference for the fluxes, and the distinctions between mass and molar units.

In September of 1957, I was asked to take part in a meeting at Purdue University for showing deans of engineering what the latest ideas were on teaching and research in a number of fields. My assignment was to discuss the teaching of mass, momentum, and energy transfer in fluid systems. My talk was, in fact, a "sales talk" for a course in transport phenomena. It was in this talk that I outlined a one-semester undergraduate course that would cover the main ideas of the subject according to the following diagram:

	MOMENTUM	ENERGY	MASS
Molecular	Viscosity	Thermal cond.	Diffusivity
Microscopic	Equation of motion	Equation of energy	Equation of species diffusion
Macroscopic	Macroscopic momentum balance & macroscopic mechanical energy balance	Macroscopic energy balance	Macroscopic species mass balances

The above presentation appeared in *Recent Advances in the Engineering Sciences: Their Impact on Engineering Education* (1958), pp. 155-177[6].

In the lecture to the deans, I emphasized the following points:

a. It is important that the students see *in one course* the connections between the transport of the *three entities* (momentum, energy, mass) and the transport at the *three levels* (molecular, microscopic, macroscopic)—all in the same notation. The students should be told that, more often than not, two or three of the transport phenomena appear simultaneously, both in industrial problems and in biological problems.

b. The students should understand the similarities and differences between the items in the three main columns. The physics describing the transport of the *three entities* should be made clear, and, when possible, the similarities in the mathematical descriptions should be emphasized.

c. Equally important is the connection between the *three levels* of description. The students should be told that one can go from the molecular level to the microscopic level, and also from the microscopic level to the macroscopic level. It is important to know how one gets the macroscopic mechanical energy balance from the microscopic equations (without invoking unnecessary and irrelevant thermodynamic arguments).

d. In addition, under *momentum transport,* I wanted to introduce the students to some of the non-Newtonian models for describing polymeric fluids and thick suspensions. Under *energy transport*, I wanted to make it clear how the heat capacity gets into the energy equation. And under *mass transport*, I wanted to explain carefully the role of various reference frames and the definition of diffusivity.

Somewhat earlier, it had been decided that a Department of Nuclear Engineering should be formed at the University of Wisconsin, with Bob Marshall as chairman of the planning committee. I was asked to serve on that committee. When it came time for developing a proposed curriculum, it was decided that a course in the basic principles of transport phenomena would be needed. I was asked to propose such a course to the Physical Sciences Divisional Committee, which had to approve new courses. The Physical Sciences Divisional Committee sent a note to the ChE Department with the query: "Why don't you have a course like this in your curriculum?" This forced the ChE Department to take up the matter. After considerable discussion, it was decided by a vote of 5 to 4 to institute the course, with the proviso that RBB would prepare a set of notes during the summer of 1957 so that the course could be instituted in the fall. RBB indicated his willingness to do this, and, within a matter of minutes, WES and ENL had indicated their interest in participating in a joint venture.

§4. BS&L roll up their sleeves

During the first half of August 1957, RBB wrote the text for Chapters 1 to 12 and then took off for a 2-week canoe trip in Canada. When he returned to the campus, assignments were made for all three authors to prepare the remaining 10 chapters. Then began a frantic semester of text preparation, making up problems, preparing figures, mimeographing the chapters, and meeting the classes. RBB held a two-hour session every week for instructing the faculty members on how best to teach the material. All three of us had other teaching duties and research supervising.

Clearly BS&L were "rushing in where angels fear to tread." No one of us was really properly prepared to undertake our task. RBB and ENL had been on the faculty for only four years, and Warren for just one. In the 21st Century, such an activity would be considered preposterous—a sure recipe for "academic suicide". But life was very different in the 1950s, when teaching and book-writing were encouraged. Furthermore, Olaf Hougen had been a great teacher and a very successful book-writer, and the three of us were greatly inspired by his wonderful accomplishments.

By the end of the fall semester of 1957, BS&L were exhausted, but the spring term was looming ahead. WES and ENL were going to make up more problems and illustrative examples during the spring, in order to make the book more useful for teaching; this was a demanding assignment, which they tackled energetically. Their years in industry had prepared them well for thinking up interesting and practical problems.

RBB, however, could not participate in this activity, since he had already arranged to spend the spring and summer in Delft (The Netherlands), supported by the Fulbright and Guggenheim Foundations. He was attracted there by the presence of Hans Kramers, who had been teaching transport phenomena since 1956 and who had prepared a set of lecture notes, titled *Physische Transportverschijnselen*[18]. RBB enjoyed very much his interactions with Hans, and we had many discussions about how to teach and use the material. During the spring term RBB gave the transport phenomena course and prepared a set of mimeographed notes *Transportverschijnselen in Stromende Media*[7].

During the summer of 1958, John Wiley & Sons prepared a limited edition of our book called *Notes on Transport Phenomena*[9]. Then between 1958 and 1960 we completely rewrote the text, making many improvements, and all the chapters were read aloud with all three authors present; in this way, we hoped to make the text homogeneous and smooth reading. In the fall of 1960 the book *Transport Phenomena*[9] was finally published.

We had a lot of fun telling our students about the secret messages in the book. In the preface, reading the first letters of the sentences gives: "This book is dedicated to O. A. Hougen. TTTM" (the last four letters stand for "this terminates the message"). And in the postface, reading the initial letters of the paragraphs gives "On Wisconsin"–the well-known school song.

The book went through 62 printings, and is now in its second edition (2002), with a "revised second edition" (2007). The first edition was translated into Spanish, Czech, Italian, Russian, and Chinese; the second into Spanish, Portuguese, and Chinese. All three of us went on to publish books in our research areas: In 1974, ENL published *Transport Phenomena in Living Systems*[20], which was the first book to combine the topics of biomedicine and transport phenomena. In 1977, RBB (together with Armstrong, Hassager, and Curtiss) published *Dynamics of Polymeric Liquids*[10,11], which included both continuum and molecular approaches. Finally in 2008, *Computer-Aided Modeling of Reactive Systems* by WES (and Caracotsios)[25] finally saw the light of day.

When the BSL book first appeared, it got a mixed reception. Young faculty members were generally enthusiastic, whereas older teachers were rather antagonistic. T. K. Sherwood of M.I.T., in a book review in 1961 in *Chemical Engineering Science*, wrote: "In a sense this is a dangerous book, for it is so well done that it will probably accelerate the trend towards emphases on analysis in the ChE curricula." Our feeling was that the transport phenomena course was, in fact, supposed to play the supporting role of an intermediate physics course that would enable chemical engineering teachers to do a better job of teaching unit operations, process control, kinetics and catalysis, and other applied subjects. We also felt that it could provide useful background for teaching courses in such diverse areas as mechanical engineering, metallurgy, soil mechanics, meteorology, and analytical chemistry.

At the same time that the department instituted the lecture course, we also established a laboratory course. For this E. J. Crosby developed a very fine set of laboratory experiments, as well as a laboratory manual, titled *Experiments in Transport Phenomena* (1961)[14]. He wanted the experiments to be valuable teaching tools and to be "student-proof." He included experiments on measurement of transport properties, measurement of profiles (velocity, temperature, and concentrations), determination of interphase transfer coefficients, and the use of the macroscopic balances. By actually performing the experiments, the students would then get a better physical feeling for the subject. These experiments proved to be so successful that he collaborated with a company to reproduce the experiments and make them commercially available to other universities.

The curricula in chemical and biological engineering, and also nuclear engineering, at UW still contain the chemical engineering course in transport phenomena. It wasn't long before Geiger and Poirier (both then at the University of Wisconsin) produced their textbook *Transport Phenomena in Metallurgy* (1973)[15], and this course is still being taught, although the authors are now at the University of Arizona. To date, the mechanical engineering department has not felt the need for a transport phenomena course in their curriculum, despite the fact that the second major textbook to appear on the subject (by Rohsenow and Choi[21] in 1961) originated in the mechanical engineering department at MIT).

It would not be appropriate to close this story by leaving the impression that we were the only engineering department that was working on a transport phenomena course in the late 1950s. We do know that several textbooks appeared shortly after ours:

Heat, Mass, and Momentum Transfer, Rohsenow & Choi (1961)[21]
Momentum, Heat, and Mass Transfer, Bennett & Myers (1962)[3]
Kagaku gijitsusha no tame no idô-sokudo-ron, Shirotsuka, Hirata & Murakami (1966)[24]
Idô-sokudo-ron, Kunii (1966)[19]

It is clear that the question posed at the beginning can be answered in the affirmative.

Finally, I would like to pay tribute to my two coauthors, Warren Stewart and Ed Lightfoot, whose contributions to the two editions of our book were substantial. Their industrial experience, their subsequent research programs, and their unique perspectives gave the book extra breadth and meaning. Along the way we had many heated arguments, but from these disagreements there usually emerged a better understanding of the subject. Our triumvirate had just the right mixture of scholarship, impatience, and "iconoclasticity."

1. W. L. Badger & W. L. McCabe, *Elements of Chemical Engineering*, McGraw-Hill, New York, 2nd Edition (1937)
2. H. Bednarczyk, *Österreichische Ingenieur- und Architekten-Zeitschrift,* **135**, 538-545 (1990)
3. C. O. Bennett & J. E. Myers, *Momentum, Heat, and Mass Transfer*, McGraw-Hill, New York (1962)

3a. R. B. Bird, "Viscous Heat Effects in Extrusion of Molten Plastics," *SPE Journal*, **11**, #7, 35-40 (1955).

4. R. B. Bird, "The Equations of Change and the Macroscopic Mass, Momentum, and Energy Balances," *Chem. Eng. Sci.*, **6**, 123-131 (1957).
5. R. B. Bird, "Theory of Diffusion," *Advances in Chemical Engineering*, Vol. 1, Academic Press, New York (1956), pp. 155-240; Corrigenda Vol. 2, p. 325 (1957)
6. R. B. Bird, "Mass, Momentum, and Heat Transfer: The Impact on Engineering Education," *Recent Advances in the Engineering Sciences*, McGraw-Hill, New York (1958), pp. 155-177
7. R. B. Bird, *Transportverschijnselen in Stromende Media*, Mimeographed Notes, Technische Hogeschool Delft (1958)
8. R. B. Bird, W. E. Stewart & E. N. Lightfoot, *Notes on Transport Phenomena*, John Wiley and Sons, New York (1958)
9. R. B. Bird, W. E. Stewart & E. N. Lightfoot, *Transport Phenomena*, John Wiley and Sons, New York, 1st Edition (1960), 2nd Edition (2002), Revised 2nd Edition (2007)
10. R. B. Bird, R. C. Armstrong & O. Hassager, *Dynamics of Polymeric Liquids Vol. 1, Fluid Mechanics*, John Wiley and Sons, New York, 1st Edition (1977), 2nd Edition (1987)
11. R. B. Bird, O. Hassager, R. C. Armstrong & C. F. Curtiss, *Dynamics of Polymeric Liquids Vol. 2, Kinetic Theory*, John Wiley and Sons, New York, 1st Edition (1977), 2nd Edition (1987)
12. R. C. L. Bosworth, *Transport Processes in Applied Chemistry: The Flow of Physical Properties in Chemical Reactors*, Pitman, London (1956)
13. S. Chapman & T. G. Cowling, *The Mathematical Theory of Non-Uniform Gases*, Cambridge University Press, 1st Edition (1939), 3rd Edition (1970)
14. E. J. Crosby, *Experiments in Transport Phenomena*, John Wiley and Sons, New York (1961)
15. G. H. Geiger & D. R. Poirier, *Transport Phenomena in Metallurgy*, Addison-Wesley, Reading MA (1973)
16. J. O. Hirschfelder, C. F. Curtiss & R. B. Bird, *Molecular Theory of Gases and Liquids*, John Wiley and Sons, New York (1954), Revised Edition (1964)
17. G. Jaumann, *Sitzungsberichte Akad. Wiss. Wien*, IIa, **120**, 385-530 (1911)
18. H. Kramers, *Physische Transportverschijnselen*, Mimeographed Notes, Technische Hogeschool Delft (1956)
19. D. Kunii, Ed., *Idô-sokudo-ron*, Iwanami, Tokyo (1966)

20. E. N. Lightfoot, *Transport Phenomena in Living Systems*, John Wiley and Sons, New York (1974)
21. W. M. Rohsenow & H. Y. Choi, *Heat, Mass, and Momentum Transfer*, Prentice-Hall, Englewood Cliffs, NJ (1961)
22. H. Schlichting, *Grenzschicht-Theorie*, Braun, Karlsruhe (1951)
23. T. K. Sherwood & R. L. Pigford, *Absorption and Extraction*, McGraw-Hill, New York (1952)
24. T. Shirotsuka, A. Hirata & A. Murakami, *Kagaku-gijutsusha no tame no Idô-sokudo-ron*, Omu-sha, Tokyo (1966)
25. W. E. Stewart & M. Caracotsios, *Computer-Aided Modeling of Reactive Systems*, John Wiley and Sons, New York (2008)
26. W. H. Walker, W. K. Lewis, W. H. McAdams & E. R. Gilliland, *Principles of Chemical Engineering*, McGraw-Hill, New York, 3rd Edition (1937)

131b A Quarter Century of Fun with Industrial Strength Transport Phenomena

David H. West, Dow Chemical Company, 2301 N. Brazosport Blvd. B-1608, Freeport, TX 77541-3257

In his 1961 review of the first edition of BSL's *Transport Phenomena*, John Butt described the book as "one of the most important texts to appear in the field of chemical engineering in many years". Clearly his assessment is still true 50 years later! Luckily BSL was written at a time when the digital computer was in its infancy and the authors had no choice but to focus on problems that were amenable to mathematical, rather than numerical, analysis. This choice demanded the student combine intuition, dimensional analysis, and fairly elementary mathematical techniques to develop qualitative understanding of the basic physics of a problem. It emphasized thinking more than computation. But students who studied the book well, and especially those who mastered it, were well equipped to tackle an enormous variety of basic research problems in engineering science, and also many practical problems encountered in the chemical industry. Great progress has been made since then in ever more complex problems such as, simulation of multi-phase and turbulent reacting flows, flow of complex fluids, convection and diffusion in porous media, and the role of molecular structure in the transport phenomena of macromolecules. This progress has enabled the field of transport phenomena to expand its relevance into the biological, materials, environmental, and energy sciences, and continues to make it a central component of chemical engineering education alongside thermodynamics and chemical kinetics.

Problems in the chemical industry frequently involve simultaneous transport of heat, mass, and momentum. The most important problems are often those in which a system behaves qualitatively different than expected, rather than just quantitatively different; for example, the catalyst lifetime is a few weeks instead of ten years, a reactor shuts down whenever it rains, or the product formed is completely different than expected. Fundamental knowledge of transport phenomena combined with other basic areas of engineering science is critical to solving such problems.

In this talk I will describe some interesting transport problems that I have encountered during my 25 years in the chemical industry.

131c Transport Phenomena from a Chemist's Perspective

Marcel A. Liauw, Itmc, RWTH Aachen, Worringerweg 1, Aachen, 52074, Germany

Chemistry students who are aware of the impact of transport phenomena on chemistry certainly are an asset in industry and academia alike. Alas, teaching transport phenomena in the chemistry curriculum does not appear to be too widespread. This may impede communication between chemistry-focused R&D and engineering-dominated process design in companies. It is therefore highly desirable to educate chemistry students in the respective concepts.

A promising approach is problem-based learning. It is being refined in the European project IMPULSE on advanced process design. In contrast to first-principles derivations of equations, short descriptions of puzzling observations are the starting point:

- Why does a heterogeneous catalytic reaction display an Arrhenius plot with two different slopes, one exactly half as high as the other?

- Why does a second nitration follow the first one in an aromatic substitution, while electron-density arguments predict it should not occur?

- How does a severely cooled semi-batch synthesis transfer into a reasonable continuous production?

Problems like these are solved in class, ideally supported by a laboratory course and/or exercises. Feedback is clearly positive, as the practical relevance of transport phenomena for chemistry is instantaneously obvious.

131d Integration of Cfd into the Undergraduate Chemical Engineering Curriculum

Jennifer Sinclair Curtis, University of Florida, Chemical Engineering Department, P.O. Box 116005, Gainesville, FL 32611-6005

The educational benefits associated with integrating computer-assisted learning and simulation technology into undergraduate engineering courses are great. Although computational fluid dynamics (CFD) has revolutionized research and design, its incorporation into the teaching of undergraduate transport phenomena has been limited. This lack of penetration of CFD into the undergraduate curriculum is due primarily to a deficiency of faculty with training in CFD, the lack of knowledge by faculty of the educational CFD software tools available, and the start up time associated with developing educational CFD materials. This presentation will share success stories on how various professors have seamlessly incorporated CFD into their undergraduate transport phenomena courses with very minimal in-class time required. In addition, the presenter will share her own experiences applying CFD into the first undergraduate transport course at the University of Florida (UF) using the educational software code FlowLab (ANSYS). FlowLab tutorials and CFD problem sets/solutions developed at the University of Florida will be disseminated to the session participants. Finally, the results of assessments used to determine the effectiveness of the addition of CFD to the UF undergraduate transport course will be discussed.

131e Panel Discussion

Jason M. Keith, Michigan Technological University, Department of Chemical Engineering, 1400 Townsend Drive, Houghton, MI 49931-1295

This is a panel discussion for this session.

The History of Chemical Engineering and Pedagogy: The Paradox of Tradition and Innovation

Phillip C. Wankat
Purdue University

Abstract

The Massachusetts Institute of Technology started the first US chemical engineering program six score years ago. Since that time, the chemical engineering curriculum has evolved. The latest versions of the curriculum are attempts to broaden chemical engineering to add product engineering, biology and nanotechnology to the traditional process engineering, chemistry and energy. Although there have been attempts to add flexibility, the chemical engineering curriculum remains monolithic (all students take almost identical sequences of courses) and hierarchical. Chemical engineering textbooks have tremendous staying power because authors have time to adapt to slow changes in the curriculum. Chemical engineering has been somewhat schizophrenic – chemical engineering research has covered all areas in which chemical engineers believe they can make a contribution, but departments have been notably unwilling (until recently) to expand the borders of the undergraduate curriculum.

Despite the conservatism of ChE departments, chemical engineering has been at the forefront of helping new professors learn how to teach and individual chemical engineering professors have been leaders in the push for engineering education reform. Yet, most ChE professors insist on lecturing. Examples of chemical engineering leadership in pedagogy include the Chemical Engineering Division of ASEE Summer School every five years, the Division's publication of the journal *Chemical Engineering Education,* and leadership in teaching professors how-to-teach. Individual efforts include the development of the guided design method, introducing Problem Based Learning into engineering, laboratory improvements and hands-on learning, the textbook *Teaching Engineering,* and the championing of cooperative group learning.

This paper will provide a brief history of chemical engineering programs, curricula and pedagogies.

Introduction and Early Programs

In 1888 MIT started Course X (Course refers to curriculum), which began as a mechanical engineering curriculum with time devoted to the study of chemistry, and eventually became chemical engineering [1, 2]. MIT did not claim invention of chemical engineering but noted that similar engineers were active in Europe [3]. Davies [4] starts his history of chemical engineering with the ancient Greeks and continues to the 1887 series of lectures presented by George E. Davis at the Manchester Technical School in England. These lectures, which were published over the next few years in the *Chemical Trade Journal,* are often considered the start of formal education in chemical engineering. Since this is the 100th anniversary of the American Institute of Chemical Engineers, we will generally limit our comments to the American experience and refer readers interested in the history of chemical engineering in other countries to the many fine chapters in Furter [5].

The historical role of MIT in starting chemical engineering education in the USA has been well documented [1-3, 6]. The initial Course X, founded by Lewis Mill Norton, was contained in the department of chemistry. Chemical engineering became a separate department in 1920 with Warren K. Lewis as the head. Perhaps the first American text in chemical engineering, *Elements of Fractional Distillation,* was published by MIT professor Clark Shove Robinson in 1922 as part of McGraw-Hill's International Chemical Series [7]. This was followed in 1923 by the seminal *Principles of Chemical Engineering* by William H. Walker, Warren K. Lewis and William H. McAdams [8], which laid the quantitative foundations of the discipline and utilized the concept of unit operations first recognized by George E. Davis (although not by that name) [4] and first delineated by Arthur D. Little in 1915 [1]. MIT also developed the idea of intensive practical education through a graduate level practice school, but this innovation has not spread beyond MIT [1, 9].

Although there were programs in practical industrial chemistry before 1888, MIT was the first school to use the title chemical engineering [2]. After MIT, the University of Pennsylvania introduced a four-year chemical engineering program within chemistry in 1892 although a separate department was not established until 1951 [2]. In 1894 Tulane started the third curriculum in chemical engineering followed by University of Michigan and Tufts in 1898 and the University of Illinois-Urbana Champaign in 1901 [2]. The first independent chemical engineering departments in the US apparently were the University of Wisconsin in 1905 [2] and Purdue University in 1911 [10].

Curriculum Developments

Early curricula were often cobbled together from existing industrial chemistry and mechanical engineering courses, and it was common, as was the case at MIT, to have no courses labeled as chemical engineering [2]. As programs grew professors of chemical engineering were assigned and specific courses in chemical engineering were developed.

AIChE became involved in studying the education of chemical engineers in 1919 through its committee on Chemical Engineering Education [11]. Between 1921 and 1922 the committee, chaired by Arthur D. Little, studied the programs at 78 schools that claimed to teach chemical engineering, and decided that chemical engineering was based on the unit operations and involved industrial scale chemical processes [11]. Although controversial, the report of Little's committee was approved in 1922 and a new committee chaired by H. C. Parmelee was given three years to determine which programs were satisfactory. This report, with the names of 14 acceptable programs, was given in June, 1925, and constitutes the beginning of engineering accreditation in the US [11]. The Engineers' Council for Professional Development (now ABET) was formed in 1932. Since AIChE was the only engineering society involved in accreditation at that time, the Institute requested and received special status. One of these perks, that a copy of each ChE program's self-study report was to be provided to the AIChE committee, was not removed until the March 2008 meeting of the ABET Board of Directors [12].

In 1925 AIChE recommended that 10.3% of the curriculum be devoted to chemical engineering courses. The recommended amount of engineering has increased over the years. In 1938 15 to 20 % of the curricula was expected to consist of chemical engineering courses [13] (Table 1). Currently, ABET does not spell out the percentages of chemical engineering courses but focuses on the skills required by graduates [14, 15]. However, the total engineering percentage has increased [14] (Table 1).

It is interesting to consider the historical development of curricula. The curricula for Purdue University, which has always had a fairly typical curriculum, are shown in Table 2 [10]. While chemical engineering was still part of chemistry (1907-08), there were no courses identified as chemical engineering and German was required since much of the chemistry literature was published in German (Table 2). In addition, a thesis was required for graduation. This plan of study was truly a combination of industrial chemistry and mechanical engineering. An increase in military training occurred during the First World War. After chemical engineering became a separate department, separate ChE courses appeared and the industrial chemistry courses disappeared (1923-24 in Tables 2 and 3). Although still required, the amount of German decreased. Both the 1907-08 and 1923-24 plans of study required a modest amount of biology. The other engineering courses included electrical and mechanical engineering, plus surveying. Descriptive geometry, required in 1907 was dropped by 1923. The 1923-24 plan of study had insufficient chemical engineering courses to meet the recommendations of the AIChE Parmelee committee, and Purdue plus many other schools were not on the AIChE list of approved schools.

Purdue (and most other rejected schools) worked hard to satisfy AIChE requirements [10]. Purdue's 1936-37 plan of study (Tables 2 and 3) satisfied the AIChE recommendations (Table 1) and Purdue was first accredited in 1933. The 28 to 31 credits of chemical engineering shown for 1936-37 in Table 3 include 6 credits of metallurgy, which was part of chemical engineering. Biology was no longer required (the other science is mineralogy). The German requirement had been reduced to 6 credits and by 1950 disappeared entirely. By 1965, shop, mechanical drawing, additional science and German had all been eliminated. The military requirement was made semi-optional and the humanities requirement (elective with a few constraints) was increased significantly. Chemical engineering requirements were increased to 25% of the course load. The 1965-66 curriculum is fairly close to the "four-year compromise curriculum light in chemistry" discussed in 1969 by Morgen [13]. The proposed 2010-11 curriculum shows the inclusion of biology, an increase in chemical engineering courses including more design, and a significant increase in the amount of hands-on laboratory (1 credit each of Fluids, Heat & Mass Transfer, and Reactor Engineering are for laboratory). The molecular basis of ChE is taught in ChE, which only partially compensates for the reduction in chemistry. This proposed curriculum has two ChE electives, an additional engineering elective, and a technical elective. Several options such as pharmaceutical engineering allow students to use their electives in an organized fashion. The military requirement disappeared during the Vietnam War.

Although total credits have dropped through the years (Table 2), the student work load appears to have stayed constant or increased. The amount of chemistry in the curriculum (Table 2) has decreased significantly. Shop, German, mechanical drawing, mechanics, and military have slowly been phased out of the curriculum. Although still available, few students select these courses. Biology has done a boomerang and returned to the curriculum. Chemical engineering science courses replaced practical, but less scientifically oriented courses after World War II [16]. The percentage of chemical engineering courses has steadily increased and there has been a trend to move these courses earlier in the curriculum (Table 3). Although not obvious from Table 3 because of the years selected, the amount of design has oscillated back and forth and is currently waxing. Hougen's [17] analysis of the curriculum trends at the University of Wisconsin are similar to those shown here, except that Wisconsin was often several years ahead of Purdue in making changes.

The current ChE curriculum at Purdue and most schools is extremely hierarchical. Starting with the first calculus course, Purdue has a seven semester sequence of required courses to graduation consisting of the calculus courses and differential equations which is a co-requisite for fluids which is followed by heat & mass transfer, which is a co-requisite for the first of two ChE labs. There are also several four semester sequences of ChE courses starting with mass and energy balances. Few of the other engineering programs have prerequisite requirements as strict.

A long term change not readily evident from looking at curricula is who teaches chemical engineering. Initially, there were no chemical engineers and the courses were usually taught by chemists and mechanical engineers. Once chemical engineers had graduated and were available to become professors, most of the chemical engineering professors had significant industrial experience and rarely had a Ph.D. [6]. Over the years an earned Ph.D. became a requirement and the expectation that engineering professors would have practical experience was lost. The current lack of practical understanding of industry and the practice of chemical engineering is obviously a problem in the education of undergraduate chemical engineers [18, 19]. The current interest in rewarding research makes it unlikely that this lack will be solved in the near future.

Current Curriculum Developments

There have been a number of recent efforts at national curriculum reform. The University of Texas-Austin Septenary committee did a major analysis of the curriculum in the early 1980's [20, 21]. The committee recommended the following: an overhaul of all the ChE courses to strengthen fundamentals and include computer calculations in all courses; inclusion of modern biology, economics and business courses in the curriculum; sufficient electives to allow specialization; and an overhaul of teaching methods and tools including major revisions of all the textbooks. The recommendations of the Committee to provide incentives for rewritting textbooks have been ignored, but many of the other recommendations made by the Septenary committee were adopted at Texas. The report also had some impact elsewhere. In particular, the need to integrate biology and chemistry into the curriculum has been widely understood [22, 23]. The need for options or tracks, which had been recommended previously [24], does not appear to have been widely adopted. The current University of Texas-Austin curriculum [25] differs from Purdue's

(Tables 2 and 3) by specifying humanities electives in American History and American Government, and requiring a literature course. In addition, an electrical engineering course is required and there are a total of six electives in science, technical and engineering areas compared to the four electives in these areas at Purdue. Both programs now require biology. Thus, the differences in these two curricula are rather small.

There has also been a push to focus chemical engineering education more on product engineering because the structure of the chemical industry has changed markedly. Many chemical engineers at both the bachelors and the Ph.D. levels now work for companies that are not considered to be chemical companies [19, 26-29], and the world of chemical engineering continues to expand [30]. Many more chemical engineers will work in specialty chemicals instead of commodity chemicals. These shifts will require more chemistry, in particular structure-property relationships including the use of quantum mechanical software. Graduates will need to be comfortable with producing products that function based on their micro- or nano-structure. In addition, there will be more interest and need to teach batch processing. Our examples and textbooks need to be revised to include examples from a much wider variety of industries. Some detailed examples of product design are available [28, 29]. At least from course titles, product design does not appear to have become a required course at MIT [31], Purdue (Table 3), University of Minnesota [32] or University of Texas-Austin [25]. Perhaps professors are including product design as examples in their courses.

Another current curriculum revision initiative is called the Frontiers in Chemical Engineering Education Initiative [33-36] that started with meetings in 2002. The initiative looks to: 1. Integrate biology into the curriculum, 2. Balance the diversity of research areas with a strong undergraduate core, 3. Balance applications and fundamentals, 4. Include both process and product design, and 5. Attract the best students to ChE. The initiative proposes that the organizing principles of chemical engineering are molecular transformations, systems and multiscale analysis. The new curriculum is supposed to be integrative and include the organizing principles plus laboratory experiences, examples, teaming and communication skills throughout the course sequence. Unfortunately, most popular chemical engineering textbooks are not arranged around the proposed organizing principles and little material for teaching within this curriculum is available. Although the initiative has been led from MIT, the current MIT curriculum [31] does not reflect this initiative. To be successful this initiative will have to convince professors that the changes are necessary, train professors in new pedagogy, and sponsor the development of an enormous amount of teaching material. In a related effort that was started independently, the chemical engineering professors at the University of Pittsburgh appear to have been convinced that these changes are necessary since Pitt has instituted a "Pillars of Chemical Engineering" curriculum [36-39]. The six "Pillar" courses on Foundations, Thermodynamics, Transport, Reactive Processes, Systems & Dynamics, and Design are block scheduled to provide additional time. The courses include molecular insight and modeling, product design, multiscale analysis, and a significant amount of simulations. Preliminary assessment data with concept maps and concept inventories shows that students are learning concepts better with the new curriculum [38, 39].

Textbooks and Other Teaching Materials

"The very boundaries of what we mean by chemical engineering are determined to a significant extent by the textbooks. The publication of "Principles of Chemical Engineering" by Walker, Lewis, and McAdams ...shaped the field of chemical engineering for many decades afterwards." [40, p. 185] In addition to Walker, Lewis and McAdams [8] Professor Bird [40] cited the books by Hougen and Watson [41], and Hougen, Watson and Ragatz [42, 43] as particularly influential. We can certainly add Badger and McCabe [44] and many other books to this list. The McGraw-Hill series of chemical engineering books started in 1925 was also very important for a number of years. Although not a textbook, Perry's Handbook [45], first published in 1934 with significant contributions from DuPont chemical engineers, has also been quite influential in chemical engineering education.

Textbooks can both constrain and open a discipline [21]. For example, BSL [46] clearly helped open chemical engineering to a more scientific approach, but later helped constrain the discipline to a continuum approach. Extremely popular textbooks such as Felder and Rousseau [47] and Fogler [48] serve to standardize parts of the ChE curriculum across the country since the vast majority of students have used these books. Because they are so widely used, the popular books can enhance or impede curriculum changes depending on the interests of the authors.

One of the current problems in chemical engineering education is, with very few exceptions, there are no young textbook authors. The first edition of most of the current ChE textbooks were written when the author(s) were in their 40s or 50s, and many of these texts are in the 2nd, 3rd, or higher editions. Younger professors are more likely to be trained in new content that should be worked into the curriculum. Unfortunately, standard advice for untenured professors is to *not* write a textbook [21, 40, 49, 50]. Professor Bird also advises, "Book-writing should not be undertaken to gain fame and fortune." [40] Although a successful textbook can pay for the college education of the author's children, the other rewards are seldom commensurate with the effort required to write a good book [40, 50]. Most chemical engineering professors are not trained in pedagogy and a really good textbook has to be based on sound learning principles in addition to being technically correct. The soundness of the pedagogical approaches is one reason for the successes of Felder and Rousseau [47] and Fogler [48]. Training all professors how-to-teach [49] would reduce the amount of on-the-job-training in writing textbooks. There have been calls for more rewards for writers of textbooks [21, 35, 40], but so far action has been sparse.

There have been attempts to use other materials besides textbooks for presenting teaching material. In the 1980's AIChE developed a series of six volumes of Modular Instruction (AIChEMI) under the overall direction of Prof. E. J. Henley. The six volumes covered Kinetics, Mass and Energy Balances, Process Control, Stagewise and Mass Transfer Operations, Thermodynamics and Transport. Modules had the advantage that the effort to write a module was orders of magnitude less than writing a textbook. Unfortunately, the quality was erratic and the modules were not widely adopted. The effort has apparently disappeared since none of the modules appears in the current AIChE catalog.

Computer aided instruction and educational games have enormous potential for improving technical education [50-53] particularly for students in the gamer generation [52]. Some of the leading ChE textbooks [e.g., 47, 48] provide supplemental instructional software as either a CD bundled with the textbook or as a course web page. Unfortunately, students often do not use the supplemental material even when required to do so [54]. Instructional games have considerable promise [53], but with current technology developing a professional quality educational game takes an order of magnitude or more effort than producing a textbook. The chemical engineering market is not large enough to support these efforts without subsidies. A major reduction in the time and cost required to develop instructional games is necessary before educational games can become economically viable to teach chemical engineering material. However, chemical engineering students may use these methods to learn calculus, chemistry [53], physics, biology, economics, and other large enrollment subjects.

Pedagogical Developments in Chemical Engineering

Since the other presentations in this symposium will discuss teaching methods in detail, I will only briefly highlight teaching methods and the contributions of ChE professors to improve teaching. Similar to all fields [50], ChE professors lecture much of the time in class. Their teaching would improve if they heeded the oft-given advice, *"Lecture less."* Instead of lecturing they could use various active and inductive learning methods such as cooperative group learning, "clickers," guided design, problem based learning, quizzes, laboratory improvements and hands-on learning, and computer simulations for part or all of the class periods [50, 55-66]. Chemical engineering professors have also been at the forefront of activities to make ABET requirements for assessment more meaningful [67, 68]. A paradox is that chemical engineering professors such as John Falconer, Rich Felder, Ron Miller, Mike Prince, Joe Shaeiwitz, Jim Stice, Charlie Wales, Phil Wankat, Don Woods, Karl Smith (an honorary ChE since his BS and MS degrees were in process metallurgy), and the entire ChE faculty at Rowan University have been at the forefront of developing and popularizing these techniques, but most ChE professors do not use them.

Chemical engineers have also been at the forefront of helping professors learn how-to-teach [49, 69-70]. The once every five years Chemical Engineering Summer School has included a how-to-teach workshop since 1987, and the popular and successful ASEE National Effective Teaching Institutes are led by chemical engineers. In addition, the Chemical Engineering Division of ASEE publishes the highly respected journal *Chemical Engineering Education* which covers new chemical engineering content and how to improve teaching and learning in chemical engineering. Teaching interested attendees to be better teachers is effective [69, 70] and it is relatively easy. Yet, it is doubtful that the majority of ChE professors have attended a formal teaching workshop or teaching course. In the past teaching workshops and courses for engineering professors were not readily available, and the reward structure at most universities did not strongly encourage faculty to improve their teaching. In my opinion *the single most effective action that can be taken to improve engineering education is to require all new engineering professors and encourage current engineering professors to take a course in how-to-teach.*

Research in improving engineering education has very recently become much more popular. This is signaled by the increased attention paid to this research by ASEE and the National Academy of Engineering, the tightening of publication requirements by the *Journal of Engineering Education* [71], the emergence of engineering education as a separate research field [72], and the development of new engineering education Ph.D. programs [73]. Chemical engineers have been at the forefront of many of these efforts. Because most engineering professors are not trained to do rigorous educational research, NSF has sponsored workshops to help interested professors start learning how to do rigorous educational research [74].

Closure

Chemical engineers active in improving engineering education are often asked why chemical engineering, which is not one of the larger engineering disciplines, has had a large impact on engineering education. I will close by speculating on the answer. Chemical engineers are interested in processes while most engineering disciplines have focused on products. Teaching and learning are processes. Thus, it is natural that chemical engineers would contribute to improving these processes. The other major engineering field interested in processes, albeit of a different type, is industrial engineering. Industrial engineering has been at the forefront of graduating Ph.D.s who did their research on engineering education. I believe that their interest in processes is a major reason that chemical engineers have been and will continue to be leaders in engineering education.

References

1. Weber, H. C., "The Improbable Achievement: Chemical Engineering at MIT," in Furter, W. F. (Ed.), *History of Chemical Engineering,* Washington, D.C., American Chemical Society, *Advances in Chemistry Series*, **190**, 77-96 (1980)
2. Westwater, J. W., "The Beginnings of Chemical Engineering Education in the United States," in Furter, W. F. (Ed.), *History of Chemical Engineering,* Washington, D.C., American Chemical Society, *Advances in Chemistry Series*, **190**, 140-152 (1980)
3. Van Antwerpen, F. J., "The Origins of Chemical Engineering," in Furter, W. F. (Ed.), *History of Chemical Engineering,* Washington, D.C., American Chemical Society, *Advances in Chemistry Series*, **190**, 1-14 (1980)
4. Davies, J. T., "Chemical Engineering: How Did it Begin and Develop?" in Furter, W. F. (Ed.), *History of Chemical Engineering,* Washington, D.C., American Chemical Society, *Advances in Chemistry Series*, **190**, 15-43 (1980)
5. Furter, W. F. (Ed.), *History of Chemical Engineering,* Washington, D.C., American Chemical Society, *Advances in Chemistry Series*, **190**, (1980)
6. Williams, G. C. and J. E. Vivian, "Pioneers in Chemical Engineering at MIT," in Furter, W. F. (Ed.), *History of Chemical Engineering,* Washington, D.C., American Chemical Society, *Advances in Chemistry Series*, **190**, 113-128 (1980)
7. Robinson, C. S., *Elements of Fractional Distillation,* McGraw-Hill, New York, 1922.
8. Walker, W. H., W. K. Lewis and W. H. McAdams, *Principles of Chemical Engineering,* McGraw-Hill, New York, 1923.
9. Johnston, B. S., T. A. Meadowcroft, A. J. Franz, and T. A. Hatton, "The MIT Practice School," *Chem. Engr. Educ., 28* (1), 38 (Winter 1994).
10. Peppas, N. A., *History of the School of Chemical Engineering of Purdue University,* West Lafayette, IN, School of Chemical Engineering, Purdue University, 1986.
11. Reynolds, T. S., *75 Years of Progress – a History of the American Institute of Chemical Engineers 1908-1983,* New York, AIChE, 1983.
12. ABET Report of the Board, http://www.abet.org/board_report.shtml Accessed June 20, 2008.
13. Morgen, R. A., "The Chemistry-Chemical Engineering Merry-Go-Round," *Chem. Engr. Educ., 3* (4), 228 (Fall 1969).
14. http://abet.org/, ABET (2007), "Criteria for Accrediting Engineering Programs. Effective for Evaluations During the 2008-2009 Accreditation Cycle," accessed 28 April, 2008.
15. Rugarcia, A., R. M. Felder, D. R. Woods, and J. E. Stice, "The Future of Engineering Education. Part 1. A Vision for a New Century," *Chem. Engr. Educ., 34* (1), 16 (Winter 2000).

16. Seely, B. A., "The Other Re-engineering of Engineering Education, 1900-1965," *J. Engr. Educ., 88* (3), 285 (July 1999).
17. Hougen, O. A., "Seven Decades of Chemical Engineering," *Chem. Engr. Prog.,* 73 (1), 89 (January 1977).
18. Landau, R., "The Chemical Engineer – Today and Tomorrow," *Chem. Engr. Prog., 68* (6), 9 (June 1972).
19. Shinnar, R., "The Future of Chemical Engineering," *Chem. Engr. Prog., 87* (9), 80 (Sept. 1991).
20. Groppe, H. (Chair), A Report by The Septenary Committee on Chemical Engineering Education for the Future, "Chemical Engineering Education for the Future," Sponsored by Department of Chemical Engineering, University of Texas-Austin, Edited by J. R. Brock and H. F. Rase (1985).
21. Sciance, C. T., "Chemical Engineering in the Future," *Chem. Engr. Educ.,* (4), 12 (Winter 1987).
22. Westmoreland, P. R., "Chemistry and Life Sciences in a New Vision of Chemical Engineering," *Chem. Engr. Educ., 35* (4), 248 (Fall 2001).
23. Mosto, P., M. Savelski, S. H. Farrell, and G. B. Hecht, "Future of Chemical Engineering: Integrating Biology into the Undergraduate ChE Curriculum," *Chem. Engr. Educ., 41* (1), 43 (Winter 2007).
24. Felder, R. M., "The Future ChE Curriculum. Must One Size Fit All?" *Chem. Engr. Educ., 21* (2), 74 (Spring 1987).
25. Chemical Engineering 2006-2008 Catalog, University Texas-Austin, http://www.utexas.edu/student/registrar/catalogs/ug06-08/ch06/courses/ch0602che-ce.html Accessed May 27,2008.
26. Cussler, E. L., D. W. Savage, A. P. J. Middelberg, and M. Kind, "Refocusing Chemical Engineering, *Chem. Engr. Prog., 98* (1), 26S (January 2002).
27. Cussler, E. L. and J. Wei, "Chemical Product Engineering," *AIChE J., 49*, 1072-1075 (2003).
28. Cussler, E. L. and G. D. Moggridge, *Chemical Product Design*, Cambridge University Press, 2001.
29. Ng, K. M., R. Gani, and K. Dam-Johansen (Eds.), *Chemical Product Design. Towards a Perspective through Case Studies,* Elsevier, Amsterdam, 2007.
30. Furusaki, S., J. Garside, and L. S. Fan (Eds.), *The Expanding World of Chemical Engineering,* 2nd Edition, Taylor & Francis, New York, 2002.
31. Massachusetts Institute of Technology, Chemical Engineering curriculum, http://web.mit.edu/afs/athena.mit.edu/org/c/catalogue/degre.engin.ch10.shtml, Accessed June 18, 2008.
32.University of Minnesota, Chemical Engineering curriculum, http://www.cems.umn.edu/academics/chen/index.php Click on "Typical Plan for 2011 or later grads (pdf)" Accessed June 18, 2008.
33. Cobb, J. T. Jr., G. K. Patterson, and S. R. Wickramasinghe, "The Future of Chemical Engineering – An Educational Perspective," *Chem. Eng. Prog., 103* (1), 30S (January 2007).
34. Armstrong, R. C., "A Vision of the Curriculum of the Future," *Chem. Engr. Educ., 40* (2), 104 (Spring 2006).
35. Frontiers in Chemical Engineering Education Initiative, http://mit.edu/che-curriculum/ Accessed May 27, 2008.
36. "Pillars of Chemical Engineering: A Block-Scheduled Engineering Curriculum," University of Pittsburgh, http://granular.che.pitt.edu/curriculum/, Accessed July 2, 2008.

37. McCarthy, J. J. and R. S. Parker, "Pillars of Chemical Engineering: A Block-Scheduled Curriculum," *Chem. Engr. Educ., 38* (4) 292 (Fall 2004).
38. McCarthy, J. J., A. A. Abatan, R. S. Parker and M. Besterfield-Sacre, "Work In Progress: Pillars of Chemical Engineering," Proceedings ASEE/IEEE Frontiers in Education Conference, Indianapolis, IN, Session F3H (Oct. 2005).
39. McCarthy, J. J., R. S. Parker and M. Besterfield-Sacre, "Results and Dissemination Plans of the Pillars Curriculum at Pittsburgh," ASEE Meeting, Philadelphia (June 2008).
40. Bird, R. B., "Book Writing and Chemical Engineering Education. Rites, Rewards, and Responsibilities," *Chem. Engr. Educ., 17* (4), 184 (Fall 1983).
41. Hougen, O. A. and K. M. Watson, *Kinetics and Catalysis,* Wiley, New York, (1947).
42. Hougen, O. A., K. M. Watson, and R. A. Ragatz, *Material and Energy Balances,* Wiley, New York, (1954).
43. Hougen, O. A., K. M. Watson, and R. A. Ragatz, *Thermodynamics,* Wiley, New York, (1959).
44. Badger, W. L. and W. L. McCabe, *Elements of Chemical Engineering,* McGraw-Hill, New York (1931).
45. Perry, J. H. (Editor-in-Chief), *Chemical Engineers' Handbook,* McGraw-Hill, New York (1934).
46. Bird, R. B., W. E. Stewart, and E. N. Lightfoot, *Transport Phenomena*, Wiley, New York (1960).
47. Felder, R. M. and R. W. Rousseau, *Elementary Principles of Chemical Processes,* 3rd update edition, Wiley, New York (2004)
48. Fogler, H. S., *Elements of Chemical Reaction Engineering,* 4th Edition, Prentice-Hall (2005).
49. Wankat, P.C. and Oreovicz, F.S., *Teaching Engineering,* McGraw-Hill, NY (1993). Available free https://engineering.purdue.edu/ChE/AboutUs/Publications/TeachingEng/index.html
50. Wankat, P. C., *The Effective, Efficient Professor. Teaching, Scholarship and Service*, Allyn & Bacon, Boston (2002).
51. Kadiyala, M. and B. L. Crynes, "A Review of Literature on Effectiveness of Use of Information Technology in Education," *J. Engr. Educ., 89,* 177 (2000).
52. Beck, J. C. and M. Wade, *Got Game. How the Gamer Generation is Reshaping Business Forever,* Harvard Business School Press, Boston, 2004.
53. Rovner, L. L., "Video Game Aims to Engage Students," *Chem. & Engr. News,* 84, p. 76, (April 10, 2006).
54. Roskowski, A. M., R. M. Felder, and L. G. Bullard, "Student Use (and Non-Use) of Instructional Software," *J. SMET Education, 2* (3&4), 41 (Sept-Dec, 2001).
55. Felder, R. M., D. R. Woods, J. E. Stice, and A. Rugarcia, "The Future of Engineering Education. Part 2. Teaching Methods that Work," *Chem. Engr. Educ., 34* (1), 26 (Winter 2000).
56. Prince, M. J., "Does Active Learning Work? A Review of the Research," *J Engr. Educ., 93* (3), 223 (2004).
57. Prince, M. J. and R. M. Felder, "Inductive Teaching and Learning Methods: Definitions, Comparisons, and Research Bases," *J Engr. Educ., 9* (2), 123 (2006).
58. Newell, J. A., "Survivor: Classroom – A Method of Active Learning that Addresses Four Types of Student Motivation," *Chem. Engr. Educ., 39* (3), 228 (Summer 2005).
59. Johnson, D. W., R. T. Johnson, and K. A. Smith, "Cooperative Learning Returns to College: What Evidence is There That It Works?" *Change, 30* (4), 27 (1998).
60. Woods, D. R., *Problem-Based Learning: How to Gain the Most from PBL,* Donald R. Woods, Waterdown, Ontario (1994).
61. Wales, C. E., R. A. Stager, and T. R. Long, *Guided Engineering Design,* West Publishing Company, St. Paul, MN (1974).

62. Dahm, K. D., "Process Simulation and McCabe-Thiele Modeling: Specific Roles in the Learning Process," *Chem. Engr. Educ., 37* (2) 132 (Spring 2003).
63. Wankat, P. C., "Using a Commercial Simulator to Teach Sorption Separations," *Chem. Engr Educ., 40*, 165-172 (2006).
64. Falconer, J. L., "Conceptests for a Thermodynamics Course," *Chem. Engr. Educ., 41* (2), 107 (Spring 2007).
65. Dahm, K. D., R. P. Hesketh, and M. J. Savelski, "Micromixing Experiments in the Introductory Chemical Reaction Engineering Course," *Chem. Engr. Educ., 39* (2), 94 (Spring 2005).
66. Farrell, S., M. J. Savelski and R. Hesketh, "Energy Balances on the Human Body: A Hands-on Exploration of Heat, Work, and Power," *Chem. Engr. Educ., 39* (1), 30 (Winter 2005).
67. Olds, B. M., B. M. Moskal, and R. M. Miller, "Assessment in Engineering Education: Evolution, Approaches and Future Collaborations," *J. Engr. Educ., 94* (1), 13 (Jan. 2005).
68. Shaeiwitz, J. A., "Teaching Design by Integration throughout the Curriculum and Assessing the Curriculum using Design Projects," *International Journal of Engineering Education*, *17*, 479 (2001).
69. Stice, J. E., R. M. Felder, D. R. Woods, and A. Rugarcia, "The Future of Engineering Education: Part 4, Learning How to Teach," *Chem. Engr. Educ., 34* (2), 118 (Spring 2000).
70. Wankat, P. C., and F. S. Oreovicz, "Teaching Prospective Engineering Faculty How To Teach," *Intl. J. Engr. Educ., 21* (5) 925-930 (2005).
71. Lohmann, J. R., "Editor's Page: Refining our Focus," *J. Eng. Educ.,* **97** (1), 1 (2008).
72. Haghighi, K., K. A. Smith, B. M. Olds, N. Fortenberry, and S. Bond, "Guest Editorial: The Time is Now: Are We Ready for Our Role?" *J. Engr. Educ., 97* (2), 119 (April 2008).
73. Wankat, P. C., Pedagogical Training and Research in Engineering Education," *Chem. Engr Educ., 42* (4), (Fall 2008).
74. Streveler, R. A. and K. A. Smith, "Guest Editorial: Conducting Rigorous Research in Engineering Education," *J. Eng. Educ.,* **95** (2) 103 (2006).

Table 1. Accreditation Recommended % in ChE Curricula [13, 14]

Topic	AIChE 1938 [13]		Topic	ABET 2008-2009 [14]
Chemistry	25-30%		Math & Basic Science	25% minimum Sufficient material to be consistent with objectives
Math	12%			
Physics	8%			
Other Sciences	2%			
Mechanics	6%			
Chemical Engineering	20-15%		Engineering	37.5% Must include design & sufficient material to be consistent with objectives
Other Engineering	12%			
Cultural Subjects	15%		General Education	Complement other components & consistent with objectives
Total	~148 credits			~124 or more credits

Table 2. ChE Plans of Study at Purdue University [10].

Topic	1907-08	1923-24[2]	1936-37[3]	1965-66	Proposed 2009-10
Chemistry	15.1%	23.7-29.9%	24.2-26.9%	16.7%	14.5%
Math	16.8%	12.3%	11.8%	12.5%	14.5%
Physics	6.6%	4.9%	5.3%	8.3%	5.3%
Other Science	1.0%	1.2-3.1%	2.0%	-----	2.3%
Mech. Draw	3.0%	2.5%	2.6%	------	-----
Mechanics	4.4%	4.9%	7.9%	2.1%	------
Ind. chem/tech	11.0%	----	---	---	-----
Chem Engr.	----	6.7-10.4%	18.3-20.3%	25.-25.7%	36.6%
Other Engr.	12.6%	12.3-19.0%	5.2%	8.3%	5.3%
Shop	7.0%	2.5%	2.6%	-------	-----
Tech electives	----	----	-----	4.9-5.6%	2.3%
Military	3.0%	3.9-13.1%	4.4%	0-5.6%	-----
English/speech	5.6%	3.7%	5.9%	3.5%	5.3%
German	10.0%	7.4-9.2%	3.9%	----	----
Other humanities	3.8%	5.5%	2.0%	12.5%	13.7%
Other	---	---	2.0%	5.6-0%	-----
Total Credits	398.5 pts[1]	163[3]-169 cr	152.7[4]-154.7	144 cr	131 cr

[1] 1 point for each hour per week in courses with no outside work and 2.5 points for each hour per week in courses with outside work. [2] Depends on options chosen. The 163 minimum was used to determine %. [3] Depends on options. The 152.7 minimum was used to determine %.

Table 3. Chemical Engineering Courses at Purdue University [10]

Semester	1907-08	1923-24	1936-37[1]	1965-66	Proposed 2010-11
1	None	None	None	None	None
2	None	None	ChE/Met, 3. (Optional)	None	None
3	None	None	None	ChE Calc, 3	ChE Calc, 3
4	None	None	None	Intro Chem Proc Ind., 3	Thermo, 4 Stat Model, 3
5	None	None	None	Thermo, 3. Fluids & Heat Trans,4	Separation, 3 Fluids, 4
6	None	Thermo, 3cr	Thermo, 3. Elem. Unit Ops, 2	Mass Transfer, 4 ChE Lab, 2	Heat/Mass Transfer, 4 Rx Eng, 4 Molec Eng, 3
7	Indus. Chem & Tech Anal, 22 points	Elements ChE I, 3. Metallurgy, 3 (Optional)	Elem. Unit Ops, 2 Unit Ops, 3 Non-Ferrous Metallurgy, 3 Pyrometry, 2 Plant Des, 2 ChE Prob, 1	Rx Kinet, 3 ChE Lab, 2 Prof. Guid. & Inspection Trips, 1 ChE Elec 3-4	ChE Lab I, 3 Proc. Dynam. & Control, 3 Des & Cost Anal., 3 ChE Elec., 3
8	Indus. Chem & Tech Anal, 22 points	Elements ChE II, 3. Metallurgy, 3 (Optional)	Inorg & Org Techn & Stoich, 3 Unit Ops, 3 Ferrous Metall., 3 ChE Prob., 1	Proc. Dynam & Control, 3 Proc. Des & Economics, 3 ChE Elec., 3	ChE Lab II, 3 Proc. Des, 2 ChE Elec., 3
Total	44 points	9-15 cr.	28-31 cr.	36-37 cr.	48 cr.

[1] Shown for the General Chemical Engineering program (Other options were Gas Technology, Metallurgy, Military, and Organic Technology).

191b The Future of Chemical Engineering and Pedagogy: The Paradox of Irresistible Forces and Immovable Professors

Richard M. Felder, Chemical and Biomolecular Engineering, North Carolina State University, Raleigh, NC 27695-7905

Several modern trends call into question the effectiveness of how we prepare students to be chemical engineers. Thanks to globalization of the economy and incredibly rapid advances in technology, most things American engineers used to do will be done in the future by either engineers in developing countries or computers. Our graduates will consequently need a whole new range of skills (critical and creative thinking, entrepreneurship, self-directed learning, knowledge of other languages and cultures,...) to get and keep jobs. Traditional engineering education was not designed to equip students with those skills, and it does a lousy job of it. It also doesn't do well at equipping students with those other unfamiliar skills ABET says they'll need if we want to be accredited—ethics, lifelong learning, and so on.

On top of all that, many axioms accepted unquestioningly by traditional engineering professors just don't stand up to hard scrutiny. You know what they are: students learn something useful by watching professors derive equations...they won't learn if you let them cooperate on homework...you've got to drill them in the fundamentals of math and science for years before you can introduce them to engineering practice...and engineering practice equals engineering design, and you can teach the students everything they need to know about it in a one-semester capstone course, and it's not necessary or even desirable for those who teach it to have ever actually done it themselves...and it's also not necessary or desirable to teach professors anything about how to teach. Those beliefs, which have been at the heart of engineering education for at least half a century and in some cases much longer, are simply wrong!

What it comes down to is that the traditional way of teaching chemical engineering—mostly straight lectures, a couple of cookbook labs, individual homework, teaching math and pure science and engineering science for three or four years and putting a semester of design at the end, and having a faculty composed almost entirely of people who have neither practiced engineering nor received training in teaching—is both inferior at promoting learning and unsuitable to produce the types of graduates who will be needed in tomorrow's job market. It's not that we don't know how to do it better—we do. We have teaching methods that have been shown by solid research to lead to better learning than the old ways do for every conceivable learning outcome, and instructional development programs that can quickly help faculty members learn to use those methods. And yet—ABET notwithstanding—mainstream chemical engineering education still has both feet firmly planted in the late 1960s, except that our students have much larger and faster and more precise and expensive slide rules to solve the same equations we solved back then and much less understanding of what the solutions mean.

In this talk, we'll review the choices our academic community faces as professors and as a profession—the conflicts between traditional and alternative ways of designing curricula, teaching, hiring faculty members, and preparing them to teach—and make some predictions about how the conflicts are likely to be resolved.

T6 #191 200 Years of Chemical Engineering Pedagogy

Pedagogy and some factors that influence how we facilitate student learning

Donald R. Woods
McMaster University
Hamilton Ontario Canada

In this paper I'll share a personal overview of pedagogy in general. This will include a listing of my personal choices of the major events, from about 1950 onwards, that have had an impact on how I try to facilitate learning. The events occurred in Canada, the United States, UK and Australia. Some were publications; some were local, provincial/state and national initiatives; some were organizations; some were networking. I'll suggest the implications and usefulness of each.

I'll also overview idealized career paths of persons who a) neglected pedagogy and focused on subject-discipline research; b) implemented pedagogy and did subject-discipline research; c) implemented pedagogy; d) implemented pedagogy and did research in both pedagogy and subject-discipline research and e) implemented pedagogy and focused on pedagogy-based research.

From this personal overview of activities in different countries, the evolution of a rich set of pedagogical ideas and a brief look at career paths of persons using pedagogy in different ways, some suggestions are given about personal actions you might take.

How might we facilitate student learning? In the 1960's when I joined academia, I thought I taught students via lectures with 50 minutes of teach talk. Yes, I tried to make the lectures interesting, entertaining and motivating. Then I joined ASEE and began to realize that many new approaches were being developed to improve student learning. In this paper I'll note the publications and major events that influenced me in my journey to improve student learning. Then I'll shift gears and give an overview of idealized career paths of persons who took different approaches to the teaching dimension of academia. Finally, I'll offer suggestions about personal actions one might take based on these two perspectives of an overview of pedagogy and options people take in their career paths.

1. Pedagogy from a personal perspective

The documentation of this journey is personal. I may have missed major pedagogical events and I may highlight ones that others might find trivial. Some of these may no longer have an impact but was pedagogy developed that provided, for me, important ideas at that time. An * indicates what I consider to be a resource that should be read today or be on your bookshelf.

1. the publication of **Bloom's taxonomy** (1). This taxonomy is a structured list representing increasing level of difficulty in learning in the cognitive domain. This has been revised by Anderson et al. (2). Such a classification is extremely helpful in analyzing the degree of difficulty expected in a task. For example, on an exam students should be given a chance to demonstrate an ability to do tasks of varying levels, rather than assigning only tasks at Bloom's level 6. Similarly, students can use such a taxonomy to monitor their growth. For the affective domain, a similar taxonomy has been developed (Krathwohl et al., 3). *

2. **McKeachie's** book on **"Teaching Tips"** (4). McKeachie provided the basics for all new teachers (and continues to provide ideas for experienced teachers). The current edition continues to provide great insight on just about any topic.*

3. **Annual workshops on pedagogy at ASEE** meetings by such people as Lois Greenfield, Gus Root, Helen Plants and Jim Stice. In the 1960s, only a few sessions related to pedagogy were offered by the AIChE. Now that has changed. Indeed, if we want to interest those faculty whose major concern is research in chemical engineering, then having sessions at the AIChE conference is the way to introduce them to ideas about how to improve teaching. These research-oriented individuals are unlikely to attend ASEE. For those interested in improving student learning we can be inspired by the presentations at ASEE and AERA conferences.

4. In the late 1960s the major event was Ray Fahien's leadership with *Chemical Engineering Education*. Ray turned this journal into a major resource for those of us concerned about scholarship in teaching. Keep up-to-date by reading this important publication. *

In the 70s...

5. The McMaster Medical School's introduction of small group, self-directed, self-assessed, interdependent **PBL**, started in the late 60s with their first class of graduates in 1972. They

created their own Center for Teaching (called the Program for Educational Development) that ran frequent "Education Rounds", published in-house reports and gave in-house workshops. I was lucky that this occurred on campus, and I could learn details about their approach. But, it wasn't until 1980 when Howard Barrows and Robyn Tamblyn published "Problem-based Learning: an approach to Medical Education" (5) that others had better access to this approach. The cognitive and psychological basis is summarized by Henk Schmidt and by Norman and Schmidt (6). However, this was still limited because for McMaster's Medical School a) one admission criteria included performance in a PBL session (whereas students in most programs were admitted based on primarily on marks), b) students admitted to the medical school were usually graduates of other undergraduate programs (and were therefore three to four years more mature than our engineering students), c) the whole program was PBL so that one faculty tutor could be assigned to a group of five students (whereas in engineering one faculty member would have a class of about 30 to 100 students) and d) most of the material was developed in the context of health sciences. Hence, attractive as this pedagogical approach might be, major modifications were needed to make it effective in our engineering classrooms. Today, there are many resources to help guide its implementation into different environments (7-9).

6. In the late 60's and early 70s William Perry published his groundbreaking analysis of "Forms of Intellectual and Ethical Development in the College Years"(10). However, it took about 10 years for the impact of this research to take effect. First, the initial approach to helping students understand their "**Perry level**" required trained professional analysis of essays. It wasn't until the 80s when people like Bill Moore established the Perry Network, Dick Culver gave workshops and Bill Moore, Peggy Fitch and Joanne Gainen created easy-to-use diagnostic tests (8) that the classroom use of the Perry inventory became used more extensively. This inventory helps students (and faculty) identify the attitudes the students hold related to the teaching and learning process. For example, students with a Perry level of 2, when placed in a PBL environment react by saying "the professor isn't doing his/her job; they are not *teaching* me." An inventory more related to developing reflective judgment and critical thinking has been developed by King and Kitchener (11). See also Rich Felder's article "Meet Your Students: 7. Dave, Martha, and Roberto." *Chem. Engr. Education, 31*(2), 106-107 (Spring 1997). Three students at different levels of Perry's Model of Intellectual Development. This can be downloaded from Rich's website (38).

7. the Ontario University Program for Instructional Development in the 1970s provided **financial support** for pedagogical projects; they had annual retreats and had external evaluation. In Canada at that time this was a very rare event. The OUPID program was described by Elrick (12). At that time, we were interested in developing *process skills* (sometimes called soft, generic, procedural or higher-order thinking skills) and our research was to learn how best to teach such skills as communication, problems solving and team work. The results from our research were published (13). In terms of process skills, one of the most challenging pedagogical issue is how best to develop confidence and skill (because lecturing was ineffective) in this domain. Conger's publications in the late 60s and early 70s for the Saskatchewan Newstart Program were an excellent resource (14).

8. A major key for learning is to have well-written, published **learning goals**. Mager (15), Kibler et al. (16) and Johnson and Johnson (17) provided excellent guidelines as to how to write learning objectives. These guidelines were the basis for developing the Keller plan for Personalized System of Instruction, PSI, or "Individualized Instructional Material". As an aside, we used their ideas in creating our workshop to develop student's skill in creating learning objectives as part of self-assessment. This is part of the McMaster Problem Solving program (13). Well-written learning objectives are still a critical part of any learning activity. Johnson and Johnson (17), Kibler et al (16) or Mager (15) remain my best resources.*

9. **Alverno College's program for the eight abilities** (18). In the mid 1970's Alverno College, Milwaukee, WI, published a list of the eight abilities as outcomes for all their programs. These abilities were effective communication, analytical capability, problem solving ability, facility in forming value judgments, effective social interaction, understanding of individual/environmental relationships, understanding the contemporary world and educated responsiveness to the arts and humanities. For each ability they published six goals/learning objectives. They trained students in self-assessment and created a separate assessment office where students could demonstrate their abilities. This revolutionary approach was, and remains, unique. I was lucky enough to be hosted by Dean Austin Doherty who graciously shared materials and helped me see how I could apply some of their approaches at McMaster. Alverno created a separate program evaluation unit to evaluate the effectiveness of their approach (19). I would encourage everyone to learn as much about their program as they can by attending their workshops and reading their publications. Their work on self-assessment is superb.*

10. In the 1970s, Keller's **personalized system of instruction**/ self-directed learning was a new approach. This prompted workshops, such as Lee Harrisberger's workshops on Individualized Learning Management or Self Paced Instruction (eg in Ontario such a workshop was held at the University of Guelph 1974) (20). Background about this approach, and variations on it continue to be used, and the principles can assist the development of distance learning modules.

11. A variation of the Personalized System of Instruction was Charlie Wales and Bob Stager's publications and workshops on **Guided Design** (21). They produced a facilitated form of problem-based learning for large groups where autonomous groups of students are given written/guided tasks to do. They have created several great resources: two books for Freshman engineering courses (including an instructor's guide) and a guide for new faculty on how to facilitate student learning via Guided Design. The book is written as a guided design format. They build their learning process around an 11-step decision-making process. I was fortunate enough to participate in several of Charlie's workshops, and the published material can be used to help craft PBL activities.

12. Craig Hogan introduced me to his research on **Jungian Typology** (22). This inventory provided students with a rich understanding of individual uniqueness and their particular style in learning, deciding and interacting. This inventory is similar to the Myers Briggs Typological Inventory, MBTI, and the Kearsey Bates inventory (23). We included this as part of the MPS

units on Personal Uniqueness and on Learning Skills (13). I recommend that this be included in all programs as the first step, following Bandura's model for self-efficacy (24), in helping students develop self confidence*.

13. Robert **Karplus** (25) created workshops including activities to develop reasoning. The activities available are in the subject domains of general science, physics, biology and chemistry. Again the pedagogical underpinnings illustrate the use of active learning.

14. Jack **Lochhead conferences** (26). In the mid 1970s Jack organized a conference on teaching reasoning, problem solving and critical thinking. He brought together key psychologists and researchers in the area of cognitive thinking (including Dorothea Simon, Jill Larkin, Alan Schoenfeld, Fred Reif, Art Whimbey, John Clement and Moshe Rubinstein). Fortunately, I was included. This was a very steep learning curve for me because these researchers were using terminology and concepts that were new. It also was a great networking opportunity. I came away with reprints and ideas that provided a strong pedagogical basis for developing problem solving skills. Additional conferences were held. A recommendation is to interact with colleagues in the cognitive and behavioural sciences and base your in-class interventions on pedagogically-sound principles (and not gut feelings).

15. The creation of ***Centers for Teaching*** and the gradual introduction of internal grants to support this activity at various Universities. Some, like the one at the University of Michigan, were established very early. At McMaster University Drs Alan Blizzard and Dale Roy, of the Instructional Development Centre, helped me immensely by TV taping my class and providing *gentle* feedback, bringing excellent workshop leaders to campus and alerting me to new developments. Frequent your Center for Teaching. *

16. Various **Newsletters** were published on developing problem solving skills and teaching (the Franklin Institute Press "Problem Solving" newsletter; McMaster University "PS News", and the HERDSA newsletter).

17. The **Pfeiffer** collection of practical workshops to develop soft skills (27). This is an excellent guide for active workshops on a wide variety of topics. I consult this resource often.

18. The publications of and workshops given by David Boud, Graham Gibbs, Alan Jenkins that brought a **European and Australian perspective**. In Canada with the Commonwealth connection, we were fortunate to have visiting educators from the UK and Australia who presented workshops (28).

19. Engineering Practice Introductory Course Sequence, **EPICS**, program at Colorado School of Mines.

20. AIChE's subcommittee on **Education Projects** and the increasing number of sessions from Group 4a at the annual meetings. Jud King's leadership; Ed Eisen's annual surveys of "how to teach subject...", the practice schools. These activities may not have had much emphasis on pedagogy but they provided very useful resources.

21. The creation of the ***Annals in Engineering Education*** as a split off from Engineering Education to focus on scholarship.

For the 1980s

22. The Jossey-Bass series **"New Directions for Teaching and Learning**" (29). This excellent series is in most Centers for Teaching and provides easy access to the fundamental research in cognition and behavior upon which to base our efforts.*

23. The creation of the Canadian **3M teaching fellowship** program (1984 onwards) had an immense impact in Canada. Ten awards are given annually from among 33,000 faculty across Canada in all disciplines. The criteria are effective in-class teaching and scholarship in teaching.

24. Marshall Lih and the **NSF programs** to financially support educational activities. Again, regrettably Canada does not have such a program.

25. Edward **deBono**'s book on the *Mechanism of the Mind* provided good background material for the MPS creativity workshop (13). His *Thinking Course* and his workshops were a great resource on how to teach thinking (30).

26. **Chickering and Gamson** (31) summarized cognitive research and suggested that we can improve student learning by applying seven basics: use cooperation not competition, expect student success, have clear time on task, account for your student's different learning styles, provide prompt feedback, use active instead of passive environments, and have extensive teacher-student interaction. *

27. Felder and Silverman's **learning style inventory** (32). Rich's articles "Meet your students..." illustrate the implications (38).*

28. The **ASEE Summer Schools** initially had negligible contributions to pedagogy but recently have included more, for example Rich Felder's contributions to the Denver Summer School. Throughout the years they have been an excellent source of how to teach different topics.

29. Noel Entwistle and Paul Ramsden's work on **deep, surface and strategic learning** (33) and their development of the **Course Perceptions Questionnaire** and the **Approaches to Studying Questionnaire** *. Dr Chris Knapper, Instructional Development at the University of Waterloo and later at Queen's University, alerted me to this research and revised the inventories to North American terminology (34). Rich Felder's article "Meet Your Students: 3. Michelle, Rob, and Art." *Chem. Engr. Education, 24*(3), 130-131 (Summer 1990) describe three different approaches to learning (deep, surface, and strategic), and the conditions that induce students to take a deep approach. This can be downloaded from Rich's website (38) A new version of the Course Experience Questionnaire has been developed to include process skills. (33) *

In the 1990s

30. **Karl Smith's** workshops and publications provide the basics for the use of various types of cooperative groups (35). *

31. **Wankat and Oreovicz** published an excellent text *Teaching Engineering.* This text can be downloaded free. Consult it often (36). *

32. **Davidson and Ambrose's** book in 1994 for young faculty (37).

33. John Prados and Stan Proctor's initiative with **ABET 2000 criteria**. Sadly the Canadian Accreditation is still bean counting.

34. **Websites**: Rich Felder has an excellent website from which you can download a rich set of resources (38). * Another excellent web source is the Society for Teaching and Learning in Higher Education (Canada) **STLHE** electronic mail **forum**. Use this forum to pose questions, follow discussions and keep up-to-date (39).*

_35. In Physics, Hestenes, Wells and Swackhamer (40) developed an inventory to test a student's understanding of the concept of "force". Steif and Dantzler (41) created a concept inventory for statics. **Ron Miller**, of Colorado School of Mines, has developed three excellent concept inventories related to thermodynamics, heat transfer and fluid mechanics (42) *. Such inventories can be used to evaluate the effectiveness of various learning environments as done, for example, by Hake (43).

36. At McMaster University several ways are used to recognize an emphasis on improving student learning. These include *The McMaster Student Union annual awards* for teaching and for lifetime achievement; the **President's Awards** for educational leadership, for resource preparation and for in-class teaching, and the Teaching Wall of Fame display. The University of Guelph took the initiative in 2000 to give an **honourary DSc** for scholarly contributions in teaching and learning. They also have a Visiting Teaching Fellow program. What options does your university offer to celebrate excellence in teaching? *

37. John **Heywood's book** is a monumental summary of "Research and Development in Curriculum and Instruction in Engineering Education" (44). * Heywood surveys and critiques papers that have been published in engineering education. Keep this reference book handy for good ideas.*

38. The series of five papers on ***The Future of Chemical Engineering Education*** published in Chemical Engineering Education (45).* Papers 2 and 3 in this series are a convenient summary of ways to improve student learning.

39. The National Survey of Student Engagement, **NSSE** (46). This North American survey provides data about a) the level of academic challenge (based on mainly Bloom's taxonomy plus

length of assignment); b) active and collaborative learning, c) student-faculty interaction (includes elements of talking to faculty outside of class, receiving prompt feedback, working on committees with faculty), d) enriching educational experience and e) supportive educational experiences. Data are given for freshman and for seniors. The 95th and 5th median data are published on the web for DRU research intensity universities at three different categories (very high activity, high and doctoral). Data are also given directly to the participating institutions. The questions can be downloaded so that you could use the same questions to gather data at the course, department and faculty levels. Extensive norm data are available. *

So what?

1. Your pedagological journey will be different from mine. However, some common elements will probably include: #1 Blooms taxonomy; #6 Perry's inventory; #8 how to create learning objectives or goals; #12 and #27, learning style inventories; #15 draw on the expertise of the professionals in your *Center for Teaching*; #26 Chickering and Gamson's seven principles and #38 the *Future of Engineering Education* series of papers.

2. Base what you do on the cognitive fundamentals. I was lucky to have been invited to Lochhead's conferences. Otherwise it would have been very difficult for me. Not all of us may be this lucky. So, ideally, attend the AERA conference. Second best is to borrow the Jossey-Bass series from your *Center for Teaching*. Next, at the AIChE conference we might annually sponsor a session on "State of the Art for Learning" to which we would invite three noted researchers from cognitive or behavioral sciences to present one-hour overviews.

3. I've noted some resources that you might want to add to your bookshelf. I also think the dual perspective of US-based innovations and Canadian-based innovations is useful. Some are similar but some are not. For example, the 3M and the hon DSc are, I think, mainly Canadian stuff (that would be nice to see in the US), and we also have a rich set of workshops (either at our universities or our national STLHE conferences) that draw from the UK and Australian connections. On the other hand ABET, NSF funding, AIChE and ASEE are really strong US elements that I wish we had in Canada.

Consider now some possible career paths.

2. Possible Careers Paths

From my experience as a consultant, as a member of the Promotions and Tenure Committee, as Departmental and Program Chair, as expert witness in a law case and as reference for candidates seeking promotion at a wide variety of universities, I offer five imaginary career paths of individuals. These faculty place different emphasis on pedagogy. Although these are imaginary, they are relatively realistic snapshot of life for research-intensive universities around 2008. Michelle, Hector, Janice, David and Frances are from different Chemical Engineering Departments.

2.1 Michelle

Michelle focused on Chemical Engineering research. She tried to be a good supervisor and teacher but her emphasis was on her research. She published about 10 papers/ annum and received the most external grants of any of her colleagues. Her research papers won awards. For in-class teaching, her student course ratings were below the average but not disastrous. She never attended any workshops to improve teaching. She was described as "a good solid lecturer". She had trouble writing a Teaching Dossier but, with the help of her mentor, her Dossier was satisfactory.
Michelle was promoted to Full Professor two years in advance.

2.2 Hector

Hector's research in reaction kinetics was going well. He received good grants and some industrial sponsorship. He produced about two refereed publications per year. Hector likes to teach and is rated as one of his Department's better teachers. Active learning is something he uses in all his classes and the students respond very positively. He attends most student events throughout the year. Occasionally visiting faculty from other Universities come to talk to Hector about his teaching. Hector attends many of the seminars given by the Center for Teaching.

Hector was promoted to Full Professor on time.

2.3 Janice

Janice loved to teach; she really wants her students to learn. As soon as she was granted tenure, she ceased applying for research grants in her specialty process control and phased out her graduate students. Yes, she continued to be a member of supervisory committees and tried to keep up-to-date with developments in process control. But her focus was on being an outstanding teacher.. and outstanding she was. Her students raved about her courses, visitors came to sit in on her classes; she won numerous student teaching awards for her in-class teaching. Her skill seemed to come naturally; she rarely consulted with colleagues in the Center for Teaching nor did she attend conferences or read educational journals.

Janice remained an Associate Professor for all her career. Indeed, she was encouraged to assume a heavy teaching load because "the students love her".

2.4 David

David was a terrific performer in class. "Spell binding", "fun", "tops in entertainment and you learned too" - these are some of the student accolades. He annually won the top awards from the students. David frequents the Center for Teaching and gives many popular workshops. He publishes papers describing teaching tips, approaches he uses in the classroom and how to interest students in any topic. The paper are published in refereed journals.

In addition, David has a research group in nanotechnology. He receives good funding and usually has one Masters and one PhD student.

When David was considered, "on time", for promotion to Full Professor the committee turned down his promotion because "we normally expect ten refereed publications. David has five refereed in nanotechnology. He also has five papers in education, but in neither field - nano or education - does David has ten".

2.5 Frances

Francis loves to teach and wants to measure the effectiveness of her classroom interventions. She also is a skilled researcher and decided to apply her research skills to teaching. She selected cooperative learning and self-assessment as her two areas of specialization.

After she attended Karl Smith's workshops at an ASEE meeting, she returned to campus and immediately introduced cooperative learning into both her undergraduate and graduate courses. Her student evaluations plummeted. In consultation with the Center for Teaching she realized that, when introducing new approached, she needed to rationalize the choice to the students, and use class ombudspersons to continually monitor the quality of the teaching-learning team. Subsequently her student ratings increased dramatically. Frances is rated one of the better teachers in the Department. To evaluate the effectiveness of her methods she gathered pre and post data using Miller's concept inventories and compared her results with the performance of students in conventional lecture classes. She also gathered data from the Course Perceptions Questionnaire and from NSSE. Frances publishes about three refereed papers/annum about her research-in-teaching. She receives grants from NSF to support her educational research. Her scholarly papers have won awards, and she is frequently asked to give seminars about cooperative learning or about self-assessment.

Frances's case for promotion to Full Professorship has been delayed for two years. In discussing this with the provost the provost admitted that the P&T committee has difficulty assessing the quality of the refereeing system used by the educational journals in which she published. "We know about *Chemical Engineering Science* and about the *AIChE Journal* but how rigorous are journals like *Assessment and Evaluation in Higher Education*? The committee will reconsider your case next year.

2.6 So What?

In summary from these cases, most institutions are research orientated and know how to measure effectiveness in Chemical Engineering research. P&T committees, and administrators tend to be learning about, but remain unconvinced and uncertain about, research-in-teaching. We need to demonstrate that refereed journals, such as "Assessment and Evaluation in Higher Education" are equivalent in reviewing standards to Chemical Engineering Science, for example.

Faculty are learning that research-in-teaching requires well-designed evaluation of pedagogical interventions. In the past we have incorrectly tended to use "they liked it" and "I liked it" evaluation. We tended to “diddle around” trying different things in the classroom without evaluating their effectiveness. We published refereed papers describing what we did in the classroom, as David did, instead of measuring the effectiveness. We should apply our well-developed research skills to evaluate our approaches to teaching.

3. Summary

From this view of activities in different countries, the evolution of a rich set of pedagogical ideas and a brief look at career paths of persons placing different emphasis on pedagogy, here are my top three recommendations.

1. Personal, starts with you. **You** can have a major impact.
2. Look beyond US, beyond AIChE, ASEE and learn from what others have done. Arrange three day visit with educators in your area of specialization. Visit your Center for Teaching often.
3. Have a realistic understanding of your local P& T system; if you don’t get tenure you can’t teach.

References

1 Bloom, B.S. et al. (1956) "Taxonomy of Educational Objectives: a classification of Educational Goals, Handbook 1: Cognitive Domain" David McKay Company, New York

2. Anderson, Lorin W., Krathwohl, David R., Airasian, Peter W., Cruikshank, Kathleen A., Mayer, Richard E., Pintrich, Paul R., Raths, James and Wittrock, Merlin C. (2001) "A Taxonomy for Learning, Teaching, and assessing: A Revision of Bloom's Taxonomy of Educational Objectives", Addison Wesley Longman, Inc

3. Krathwohl, D.R. et al. (1964) "Taxonomy of Educational Objectives - the classification of educational goals, Handbook II, affective domain," David McKay, New York.

4. McKeachie, W.J. (1951) "Teaching Tips: a guidebook for the beginning College Teacher" D.C. Heath and Co., Lexington, MA and (2001) 11th edition, D.C. Heath and Co.,

5. Barrows, H.S. and Robyn Tamblyn (1980) "Problem-based Learning: an approach to medical education" Springer.

6. Schmidt, Henk G. (1983) "Problem-based Learning: rationale and description," Medical Education, 17, 11-16; Norman G. and H. Schmidt (1992) "The Psychological basis of problem-based learning: a review of evidence," Academic Medicine, 67, no. 9, 557-565

7. Boud, D. (1985) "Problem-based Learning in Education for the Professions" HERDSA, Sydney, Australia; Boud, D., and Grahame Feletti (1997) "The Challenge of Problem based Learning" 2nd ed. Kogan Page, London; Knowles, Malcolm (1975) "Self-directed learning: a guide for learners and teachers," Follett Publishing, Chicago

8. Woods, D.R. (1995) "Problem-based Learning: how to gain the most from PBL" Woods, Waterdown, Canada. This book is written for students to help them work effectively in a PBL environment. For faculty, see http://www.chemeng.mcmaster.ca/innov1.htm from which you can download three books "Preparing for PBL", "Problem-based learning: resources to gain the most from PBL" and "Problem-based learning: helping your students gain the most from PBL"

9. De Graff, Erik and Anette Kolmos (2007) "Management of Change: implementation of Problem-based Learning in Engineering" Sense Publishing, Rotterdam

10. Perry, William G. (1970) " Forms of Intellectual and Ethical Development in the college years, a scheme", Holt, Rinehart and Winston.

11. King, P. M. & Kitchener, K. S. (1994) "The development of reflective judgment in adolescence and adulthood". San Francisco: Jossey-Bass. And "Reflective Judgment Scoring Manual" from Reflective Judgment Associates.

12. The Ontario Universities Program for Instructional Development, Elrick, M. (1990). "Improving instruction in universities: A case study of the Ontario Universities Program for

Instructional Development (OUPID)", The Canadian Journal of Higher Education, 20(2), 61-79

13. Woods, D.R., et al.,(1997) "Developing Problem-solving skills: the McMaster Problem Solving Program," J. of Engineering Education, 86., 2, 75-91. For faculty, see http://www.chemeng.mcmaster.ca/innov1.htm and visit the MPS site where we are gradually posting details about the MPS units.

14. Conger, Stuart (1969 to 1973) "Life Skills Coaching Manual" Department of Manpower and Immigration, Prince Albert, Sask; Stuart Conger (1973) "Readings in Life Skills" Department of Manpower and Immigration, Prince Albert, Sask; Joan Hearn (1981) "More Life Skills" Employment and Immigration Commission, Ottawa.

15. Mager R. (1962) "Preparing Educational Objectives" Fearon Publishers, San Francisco. 3rd edition, (1997)

16. Kibler, R.J. et al. (1974) "Objectives for Instruction and Evaluation" Allyn and Bacon

17. Johnson, S.R. and R.B. Johnson (1970) "Developing Individualized Instructional Material: a self-instructional material in itself," Westinghouse Learning Corporation, New York

18. Alverno College's program for the eight abilities. Alverno College Faculty (1976) "Liberal Learning at Alverno"; (1976) Doherty, A., Riordan, T., Roth, J., (2002). *Student learning: A Central focus for institutions of higher education.* WI: Alverno College Institute.

19. Mentkowski, M. et al., (2000) "Learning that Lasts" Jossey-Bass, San Francisco

20. Grayson, L.P and J. M. Biedenbach (1974)"Individualized Instruction in Engineering Education" ASEE, Washington DC; "Keller Plan", see Billy V. Koen, Chpt 4 in Grayson and Biedenbach; Harrisberger, Lee, "The Management of Individualized Instruction Program", Chpt 8 in Grayson and Biedenbach.

21. Wales, C.E., R.A. Stager and T.R. Long (1974), "Guided Engineering Design: Project Book", West Publishing Co.; ibid, (1974) "Guided Engineering Calculations"; (1977) C.E. Wales, R.A. Stager, "Guided Design"; (1973) "Educational Systems Design"; (1974) C.E. Wales, "The Systems Approach: an introduction" Chpt 2 in Grayson and Biedenbach.

22. Hogan, R. C. And D. W. Champagne (1979) Personal Style Inventory" personal communication, 1979. This is based on Carl Jung's typology. A popular commercial version of this inventory is called the Myers Briggs Typology Inventory, MBTI. Another version is published in Keirsey and Bates' book "Please Understand Me" (23).

23. Keirsey, D and M. Bates (1984) "Please Understand Me: character and temperament types", Prometheus Nemesis Books, Del Mar, CA and http://www.keirsey.com 2008. The on-line version uses the same 70 questions as are given in the book pages 5 to 10. Your responses are scored and you may download free your "major type" from among Guardians, Idealists,

Rationals and Artisans. Keirsey includes four Jungian types in each of his four categories. These differ in name between the web results and the book. For example, the "Guardian" includes ESTJ, ISFJ, ISTJ and ESFJ. To obtain more details about the type from the web version requires payment. *Comment:* the seventy questions offer either yes or no answers whereas Hogan's version allows you to distribute a numeral from 0 and 5 between the two options (as long as the total is 5). Thus, in question 1 from Keirsey, "at a party do you (a) interact with many, including strangers or (b) interact with a few, known to you." Keirsey expects an either-or answer. I might prefer (a) 3 and (b) 2, meaning that I usually prefer meeting new people but I certainly enjoy visiting with known friends. Furthermore, Hogan's version provides a numerical value between 0 and 40. Thus a score of 20 on the NS scale would suggest that I am either N or S; whereas a 30 N, 10 S would suggest I am rather strongly N. I prefer the Hogan version.

24. Bandura, A. (1982) "Self efficacy mechanism in human agency" *The American Psychologist*, 37(2), 122 - 147. Bandura, A. (1986). *Social foundations of thought and action: A social cognitive theory*. Englewood Cliffs, NJ: Prentice Hall.

25. Karplus, Robert (1977) "Science Teaching and the Development of Reasoning: General Science" a workshop, Lawrence Hall of Science, University of California, Berkeley, CA

26. Lochhead, Jack and John Clement (1979) "Cognitive Process Instruction: research on teaching thinking skills", The Franklin Institute Press, Philadelphia; D. Tuma.and F. Reif (1980) "Problem Solving and Education: issues in teaching and research", Lawrence Erlbaum Associates, Hillsdale

27. Pfeiffer, J.W. and J.E. Jones (1979) A Handbook of Structured Experiences for Human Relations Training: vols I to VII and reference guide to handbooks and annuals, 3rd ed.," University Associates, La Jolla, CA

28. Example publications from these authorities include Gibbs, Graham (1992) "Teaching More Students: volumes 1 to 5", Oxford Brookes University, Oxford UK; Boud, David (1995) "Enhancing Learning through Self Assessment", Kogan Page, London

29. Jossey-Bass series (1980 onwards) the series "New Directions for Teaching and Learning", about 4 new volumes per year. Jossey-Bass, San Francisco, CA

30. deBono, Edward (1971), "Mechanism of the Mind" Pelican Books, Harmondsworth, Middlesex, England and (1982) "deBono's Thinking Course" BBC, UK.

31. Chickering, A.W. and Z.F. Gamson (1987) "Seven Principles for good practice in undergraduate education" AAHE Bulletin, Mar 3-7

32. Felder, R.M. and L.K. Silverman (1988) "Learning and Teaching Styles in Engineering Education," Engineering Education 78, 674-681 and subsequent articles by Rich Felder about the implications.
Stan and Nathan." *Chem. Engr. Education, 23*(2), 68-69 (Spring 1989).

--, . Susan and Glenda." *Chem. Engr. Education, 24*(1), 7-8 (Winter 1990). The sequential learner and the global learner on the Felder/Silverman learning styles model.
-- Jill and Perry."*Chem. Engr. Education, 25*(4), 196-197 (Fall 1991). The judger and the perceiver on the Myers-Briggs Type Indicator.
– Edward and Irving."*Chem. Engr. Education, 28*(1), 36-37 (Winter 1994). The extravert and the introvert on the Myers-Briggs Type Indicator
-- Tony and Frank." *Chem. Engr. Education, 29*(4), 244-245 (Fall 1995). The thinker and the feeler on the Myers-Briggs Type Indicator
These articles and the inventory can be downloaded from Felder's website (38).

33. Entwistle, N. & Ramsden, P. (1983). *Understanding student learning.* London: Groom Helm. A new version of the Course Experience Questionnaire has been developed to include process/generic skills. http://www.engsc.ac.uk/downloads/experiencequestionnaire.ppt

34. Woods, D.R. (Forthcoming) "Motivating and Rewarding University Faculty to improve Student Learning: a guide for faculty and administrators". The North American versions of Approaches to Studying Questionnaire and the Course Perceptions Questionnaire are given in the Appendix of this book.

35. Johnson, D.W., R.T. Johnson and Karl A. Smith (1991) "Active Learning: cooperation in the College Classroom," Interaction Book Co., Edina, MN

36. Wankat, P.C. and F.S. Oreovicz, (1993) *Teaching Engineering*, McGraw-Hill, NY, 1993. Available free as pdf files https://engineering.purdue.edu/ChE/AboutUs/Publications/TeachingEng/index.html

37. Davidson, C.I and Susan A. Ambrose (1994) "The New Professor's Handbook" Anker Publishing, Boston, MA

38. Felder, R.H. website http://www4.ncsu.edu/unity/lockers/users/f/felder/public/, Sept 2008

39. Society for Teaching and Learning in Higher Education (Canada) STLHE; for the forum contact <STLHE-L@UNB.ca>

40. Hestenes, D., M. Wells and G. Swackhamer (1992) "Force Concept Inventory", The Physics Teacher, 30, no. 3, p 141 -158

41. Steif, P.S. and J.A. Dantzler (2005) " A Statics Concept Inventory: development and psychometric analysis" J. Engng. Ed. 94, no. 4, 363-371

42. Ron Miller, Chemical Engineering Department, Colorado School of Mines, personal communication, 2007

43. Hake, R.R. (1998) "Interactive-engagement vs. Traditional Methods: a six thousand student survey of mechanics test data for introductory Physics courses," American J. of Physics, 66,

64-74

44. Heywood, John (2005) Engineering Education: Research and Development in Curriculum and Instruction" John Wiley, Hoboken, NJ

45. Series of five papers coauthored by R. Felder, A. Rugarcia, Jim Stice and Don Woods that appeared in Chemical Engineering Education, 2000. These can be accessed via Rich Felder's website (38). A. Rugarcia, R.M. Felder, J.E. Stice and D.R. Woods, (2000) "The Future of Engineering Education," **I** A Vision for a New Century," Chemical Engineering Education, 34, 1, 16-25; R.M. Felder, D.R. Woods, J.E. Stice and A. Rugarcia (2000) "The Future of Engineering Education," **II** Teaching Methods that work," Chemical Engineering Education, 34, 1, 26-36; D.R. Woods R.M. Felder, J.E. Stice and A. Rugarcia Torres (2000) "The Future of Engineering Education," **III** Developing Critical Skills," Chemical Engineering Education, 34, 2, 108-117; J.E. Stice, R.M. Felder, D.R. Woods, and A. Rugarcia Torres (2000) "The Future of Engineering Education," **IV** Learning How to Teach," Chemical Engineering Education, 34, 2, 118-127; R.M Felder, A. Rugarcia and J. Stice (2000) "The Future of Engineering Education," **V** Assessing Teaching Effectiveness and Educational Scholarship," *Chem. Engr. Education, 34* (3), 198–207

46. National Survey of Student Experience. Http://www.nsse.iub.edu/

191d 200 Years of Chemical Engineering Education: The Need for Both Scholarly Teaching and the Scholarship of Teaching and Learning

Michael J. Prince, Chemical Engineering, Bucknell University, Department of Chemical Engineering, Bucknell University, Lewisburg, PA 17837

Most engineering faculty teach in the way that they were taught, not because they're convinced by the research that supports the effectiveness of traditional methods but because these methods are familiar and comfortable. Curricula in chemical engineering programs have remained essentially unchanged for 50 years for much the same reason. If our pedagogy and curricula were unchanging because the best available research evidence showed that both were meeting our needs better than any alternatives, that would be fine. Unfortunately, that just isn't the case.

Clearly, engineering programs should evolve to reflect changes in what we know about teaching and learning. It's just common sense to let the best available research findings guide our decisions as educators, and this presentation will discuss at least two consequences of accepting that conclusion. First, there must be professionals who are deeply knowledgeable about engineering to help conduct the necessary research. That work of extending what we know about teaching and learning within a discipline is the scholarship of teaching and learning. The second consequence of accepting the original premise is that engineering faculty must incorporate the best available research findings when developing their classes and curricula. That activity is frequently called scholarly teaching.

This presentation will make the case that these activities, scholarly teaching and the scholarship of teaching and learning, are both essential if chemical engineering education is to evolve over the next century and be as good as it needs to be. The talk will offer examples of both scholarly teaching and the scholarship of teaching and learning, and will provide an overview of opportunities and challenges in each area as chemical engineering as a discipline moves into the next century.

Seeing Is Believing - Learning Fundamentals of Chemical Engineering

Ganapati D. Yadav,
Department of Chemical Engineering,
University Institute of Chemical Technology (autonomous),
Nathalal Parekh Road, Matunga, Mumbai, 400019,
India

Chemical Engineering has become a very pivotal discipline which enjoys commonality with all cutting edge sciences. Chemical engineers are the most versatile species who are adept at converting themselves into chemists, physicists, material scientists, biologists, biotechnologists, mathematicians, environmentalists and all sorts of engineers. They can certainly disguise themselves as one such species and still retain their identity. This has been possible due to the sound training chemical engineers receive in basic sciences, mathematics and systems approach. The importance of experiments in any scientific field can be hardly overemphasized and more the number of experiments invoking different basic principles, the better it is. Apart from the regular experiments, it is important to introduce many concepts in chemical engineering through visualization.

In 1978, M.M. Sharma, the legendary Professor and Head of Chemical Engineering at this Institute, returned from his trip to the Cambridge University UK and discussed with me a simple demonstration experiment performed by J.F. Davidson on the similarity between bubble flow in liquids and fine solids. Davidson apparently held two glass tubes in his hand side by side, each filled with water and fine particles to the same level with a little air on the top, closing the tops with his palm. He just inverted the tubes and asked Sharma to observe the bubble flow. What Sharma could not believe was that the bubble travel in gas-solid and gas-liquid systems was strikingly similar. He asked me make a simple set up to show this experiments to the UG students. As a young Associate Lecturer, a sort of demonstrator and also a Ph. D. student, I was heavily involved in Chemical Engineering Laboratory and enjoyed setting up experiments for the UG students. I enjoyed doing experiments and did all sorts of experiments before doing theory.

Immediately I made the first set-up with two glass tubes of 1 cm diameter and 1 m in height, with silica powder (~300 μm) and water-glycerol as the 'fluids' leaving 2 cm air space at the top. I fixed these tubes side-by-side on a wooden plank (10 cm diam x 1.20 m length) made a rotary device with a handle at the back and put it on a frame (Fig. 1). With a few trials, I could replicate Davidson's experiment. Only after the experiment was done, I looked at Davidson's work and tried to understand the theory! A bed of solid particles fluidized by gas behaves like a fluid. There is a minimum gas flow for the fluidization at which the pressure drop is just enough to support the weight of the bed. When the gas flows at a velocity greater than that required for minimum fluidization, each particle is suspended and the whole bed becomes mobile and resembles a liquid. The bubble does not break and maintains its fidelity and much of the gas travels as a bubble.

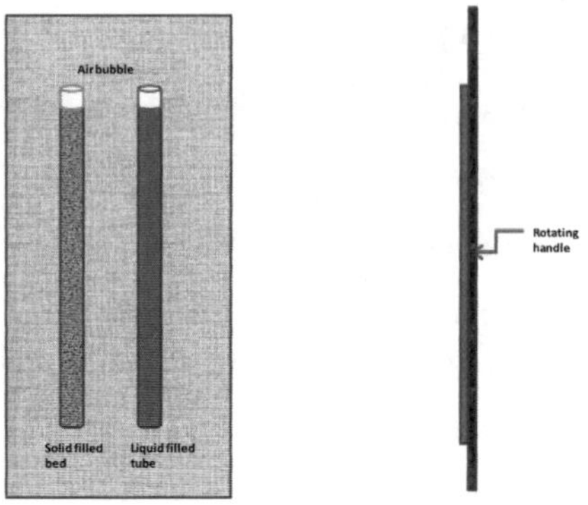

Fig. 1 : Similarity of bubble flow in gas-solid and gas-liquid systems

Professor Sharma was too excited to settle at one experiment and made a continuous effort to generate ideas as well as gave us the freedom to add experiments of such kind. Sharma wanted me and some of my other junior colleagues teaching Chemical Engineering Lab to set up a Demonstration Laboratory where all fundamentals of Chemical Engineering could be taught through simple experiments to reinforce the concepts. Being a celebrity and active consultant, Sharma thought this unique laboratory should be developed and made a regular 'habit' for the disbelieving students. He was very enthusiastic about Marangoni effect, three phase fluidization, mixing, etc. Ours was a very poorly funded lab which did not have the material and money to buy costly equipment/apparatus and we had to live on ideas and some easy to fabricate set-ups. He used to throw challenges at us and asked all research students to come with any demonstration related to their work which could become part of this lab. We have developed a special lab called 'Demonstration Experiments' which was introduced. Several concepts when visualized become very clear to the doubting students and they appreciate them greatly.

How to bring Marangoni effect to the lab and in a country, where even looking at an effervescent beer is a taboo, was a big problem ! (at least in 1970s!!). Marangoni's famous story is related to mass transfer when a bottle of champagne is opened up and the interfacial turbulence. A surface tension lowering solute on transfer from one phase to another causes imbalance in the interfacial turbulence. So we chose acetone as the surface tension lowering solute being transferred from toluene to water. A 4% v/v solution of acetone in toluene makes the continuous phase whereas water droplets are added from a burette. When the drop moves in the organic phase, it disturbs it resulting in non-uniform eddies. This brings an acetone rich element to the interface, which lowers the interfacial tension and changes the pressure of the drop and the drop 'kicks'. In the case of gas-liquid this can be demonstrated by putting a few drops of diethyl ether over water in a watch glass and sprinkling talcum powder.

Thus the next experiment on our radar was the internal circulation in droplets, which was done with carbon tetrachloride and aluminum particles to demonstrate how fluid droplets behave in different manner vis-à-vis solid particles while travelling in a liquid medium. There is onset of internal circulation because of the mobile interface. This simple experiment was done in a glass cylinder containing water and releasing organic phase droplet containing fine aluminum powder, with a light source at the background. The internal circulation leads to higher mass transfer rate in liquid-liquid systems. Now this experiment has graduated to an elaborate set-up.

Another concept was based on the direction mass transfer, coalescence and identification of continuous and dispersed phases. It is commonly believed that the phase with lesser volume is a dispersed phase. However, with a phase volume upto 70%, it can still be a dispersed phase. So a simple demonstration is to take two immiscible liquids, one of them with a dye having no distribution in the other phase and agitate it. As the agitation is stopped, the interface is distinguished after some time. The dispersed phase droplets, still coalescing, crowding and merging with its bulk phase will be seen next to the interface in that phase. The other side of the interface is free of any droplets. No great funds are required to show this behavior!! The volumes can be varied to see which would be the dispersed phase in 30: 70, 50:50 and 70:70 phase ratios. The idea can be extended to make one of the phases as the continuous or dispersed phase by locating the impeller above or below the interface. If the heavier phase is desired to be the continuous phase, then the impeller is placed in it and the lighter phase becomes dispersed. On the contrary if the impeller is placed in the lighter phase, the heavier phase is dispersed in it. However, this may be an unnatural position for the impeller. Hence if the lighter phase is required to be continuous, then placing the impeller near the bottom of the vessel, adding the lighter phase firstly, agitating it and then adding the heavier phase leads to the desired effect.

A related experiment in this regard is the phase inversion in liquid-liquid contactors, such as mechanically agitated contactors and rotating disc contactors, which is governed by several parameters such as interfacial tension, phase ratio, density, viscosity, etc. Phase inversion occurs when the rate of coalescence exceeds that of redispersion. It is difficult to disperse a high surface tension liquid at high hold up in a low surface tension liquid. An organic phase such as toluene and colored aqueous can be selected to observe phase inversion. Dispersed phase flow rate is increased till the creation of a dense layer of droplets at the bottom of the column. This inversion normally occurs at the bottom compartment and passes in upward direction from compartment to compartment.

I had initiated research on phase transfer catalysis (PTC) in mid-1970s for my doctoral thesis and this new technique was hardly known to chemical engineers. Liquid-liquid reaction between reagents in two different phases is extremely sluggish and takes days to complete. The enhancement in rates of reaction is seen due to transfer of the nucleophile from aqueous phase to organic phase, where the substrate resides, by using a phase transfer agent (Fig. 2).

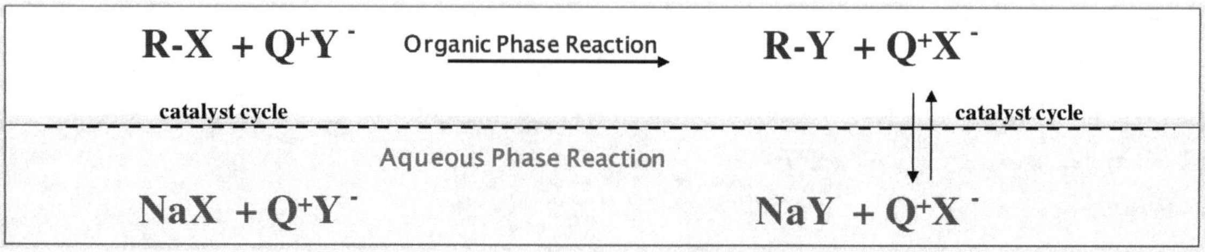

Fig. 2 : Liquid-liquid phase transfer catalysis

To demonstrate PTC, a simple experiment would be the esterification of nitrophenol with acetic anhydride in a stirred vessel. The sodium salt of nitrohenol is taken in the aqueous phase and has a yellow color. Acetic anhydride is taken in mixed xylene as a solvent and is colorless. Even after intense mixing, no reaction takes place and two layers are separated. Addition of a small quantity of tetrabutylammonium bromide as a catalyst and agitation shows that the reaction has taken place in the organic phase and there is a discoloration of the aqueous phase and coloration of the organic phase due to transfer of quaternary nitrophenoxide to the organic phase. This experiment can be done in a beaker without any agitation while observing gradual discoloration and coloration of the phase or the rates could be enhanced by agitation.

Weissenberg effect is well known to polymer chemists and engineers. The role of mixer is quite adequately demonstrated in non-Newtonian fluids where the ordinary M.A.C. will not be a good idea; for instance, viscoelastic materials which possess properties of viscous materials as well as elastic solids. Materials possessing elasticity exhibit normal stress difference. In a rotational flow, the fluid elements experience centrifugal force and elastic force. At lower rotational speeds where the elastic force exceeds the central force, the material moves towards the centre and to keep continuity climbs on the shaft. A 4% w/w aqueous polyvinyl alcohol with 3% v/v glycerol can be used as a viscoelastic material to demonstrate this ‘climbing’, known as Weissenberg effect. This demonstration can be made dramatic by taking two identical beakers without any baffles, one containing water and another the viscolelastic liquid and asking the students to predict the outcome of the agitation. One can literally trick the students and then explain the phenomenon.

Three phase catalytic reactors are so common in industry and the reactor performance is marred by poor mixing. A complete suspension of solids is very important and the operation has to be beyond the critical speed. Contrary to the common belief, when gas is introduced in an agitated solid liquid system, the particles will settle down if the speed is in the critical region. This is detrimental to many catalytic reactions such as dead end hydrogenations. This experiment is done with a fine sand particles in water in a stirred tank and the speed is increased gradually to keep the solids suspended. Then air is introduced into the tank to find that the particles settle down. The speed has to be increased to a higher value to make suspend the particles.

Table 1: Demonstration Experiments

1	Laminar flow in pipes	22	Interfacial polycondensation
2	Internal circulation	23	Foam Fractionation
3	Direction of mass transfer & coalescence	24	Hydrotropy
4	Interfacial turbulence	25	Small scale experiment with surfactants
5	Agitation in liquid-liquid systems: Identification of continuous and dispersed phases	26	Detecting reverse micelles in non-polar media
		27	Microemulsions
6	Phase Inversion in agitated contactors	28	Incompatibility of polymers (Aqueous two phase extraction)
7	Wetting characteristics	29	Segregation due to flow (coning)
8	Induced coalescence	30	Relaminarization
9	Liquid maldistribution in packed towers	31	Enhanced sedimentation in settling tanks (Boycott effect)
10	Wire mesh type of entrainment separator	32	Air lift reactor
11	Gas-inducing type of impeller	33	Experiment with granular particles
12	Co-current gas-liquid down flow in packed columns	34	Circulation cells in bubble column
13	Flooding in gas-liquid and liq-liq contactors	35	Origin of viscoelasticity in fluids
14	Bubble rise in gas-liquid and gas-solid system: A similarity	36	Oscillating reactions
15	Three phase fluidization: Contraction of liquid fluidized bed on introduction of gas.	37	Langmuir monolayer of surfactant
		38	Ultrafiltration
16	Three-phase reactor	39	Laser Doppler Anemometry
17	Weissenberg effect	40	Cavitation
18	Cyclone separator	41	Hydrogenation reactions in high pressure vessel
19	Instantaneous gas-liquid reaction: Existence and movement of reaction plane	42	Jet mixer
20	Selectivity in two-phase liquid-liquid systems		
21	Phase transfer catalysis		

A large number of these demonstration are a result of our own research in various areas and some published work of other as listed in the references. The following is a partial list of such experiments which we have developed over last 30 years is given in Table 1 .

A totally renovated lab is now set up to house all these experiments, video demonstrations, posters and portraits of great scientists and engineers to inspire the young minds. With advent of new equipment and tools, we have been striving hard to include nanotechnology, green chemistry and engineering and biotechnology experiments. Several international visitors have appreciated our efforts and now we are according the University Grants Commission's Networking Centre in Chemical Engineering to educate teachers, students and practitioners from industry on a sustainable basis. When M.M. Sharma emphasized the need of such demonstration lab during 1970s, hardly anybody would have realized its importance in chemical engineering curricula.

References:

Andew S.P.S., Acta Techn. Chimica, p153 (1961)

Anon. *Chem. Eng. Edu.* 10, 180 (1976)

Bingham E.C. and Wikoff R.W., *J. Rheol.* **2,** 395 (1931)

Davidson J.F. 'Bubble motion in liquids and fluidized beds' (movie)

Davies J.T., *Turbulence Phenomenon*, Academic Press, New York, 1972.

Erlich F.R., *Rheology Theory and Applications*, Vol. 1, Academic Press, New York, 1956.

Grouthuis H. and Zuiderweg F.J., *Chem. Eng. Sci.* **11**, 288 (1960).

Heertjes P.M. , Yan Mens M.H. and Butaye M., *Chem. Eng. Sci.* **10**, 47 (1959).

Hoffman H. and Ebert G., *Angew. Chem.*, **27**, 987 (1988).

Huber M. and Mier W., *Sulzer Technical Review*, **1**,1 (1975).

Joshi J.B. and Sharma M.M., *Can J. Chem. Eng.* **55**, 689 (1977).

J. Chem. Edu. 34, 484 (1957).

J. Chem. Edu. 48, 385 (1958).

Kinter R.C. *Adv. Chem. Eng.* Vol. 4, 1963.

Leva M., *Tower Packings and Packed Tower Design*, US Stoneware Co., Akron, Ohio, 1953.

Limlich R., *Chem. Eng.* **75**(27), 95 (1968).

Lindman B., *Soap and Water : A Complicated System*, Kem. Tidskr. No. 9 (1981).

McBain et al., *J.A.C.S.* **62,** 2880 (1940).

McBain and Merrile, *Ind. Eng. Chem.* **34**, 915 (1942).

Martin G.D., *Ind. Eng. Chem. Proc. Des. Dev.* **11**, 397 (1972).

Midouz N., Favier M. and Charpentier J.C., *J. Chem. Eng. Japan*, **9**, 350 (1976).

Morgan P.W., *Condensation Polymers by Interfacial and Solution Methods,* Academic Press, New York, 1965.

Nagata S., *Mixing Principles and Practice*, Wiley Interscience, New York, 1975.

Onde K., Kobayashi T., Fujine M. and Takahashi M., *Chem. Eng. Sci.* 26, 2009 (1971).

Pangarkar V.G., Yadav G.D. and Niranjan K., *Demonstration Experiments in Chemical Engineering*, Chemical Engineering Education Development Centre, I.I.T. Madras, 1981.

Perry R.H. & Chilton C.H.(Ed.), *Chemical Engineers' Handbook*, McGraw-Hill, New York, 18-84, 1973.

Reynolds O. *Phil. Mag.* **20**, 469 (1885).

Sarkar S., Philips C.R., Mumford C.J. and Jeffrey G.V., *Trans. Inst. Chem. Eng.* **58**, 43 (1980).

Stairmand C.J. , *Trans. Inst. Chem. Eng.* **29,** 356 (1951).

Weekman V.M. Jr. and Myers J.E., *AIChE J*, **10,** 951 (1964).

Yadav, G.D., *Ind. Eng. Chem. Res.* **46**, 2911 (2007).

Yadav, G.D., *Top. Catal.* **29**, 145 (2004).

York O.H., *Chem. Eng. Prog.*, **50**(8), 421(1954).

Ziokornik M., Chem. Eng. Tech.,**38**, 717 (1966).

Structuring a Chemical Engineering Curriculum for the Globalized Era

K. C. Khilar[1], K. P. Madhavan[2], A. Kudchadkar[1] and P. Ray[3]

1. Institute of Petroleum Technology, Gandhinagar
2 Indian Institute Of Technology, Bombay
3. Department of Chemical Engineering, Calcutta University

A Brief Review:

Establishment of Chemical Engineering as an independent discipline took a long time- from the eighteenth century to the beginning of twentieth century, and its search for "self" or "identification" continued for an additional couple of decades. As is well known, the chemical engineer had to establish himself – first by proclaiming his distinct identity and then winning recognition from the engineering, scientific and industrial community. And subsequently he had to reinvent himself - that is redefine his field of expertise, knowledge and application several times – or rather continuously over the last hundred years. Historically the "modern" chemical engineering became recognizable only in 1950's and possibly1960 's but the process of redefinition or evolution has continued unabated. The history of development of Chemical Engineering courses from Industrial Chemistry to present days "multiscale" through phases of "unit operations" , "chemical engineering science" is well known.

Chemical Engineers as a group seem to be greatly concerned with self identification and as a natural consequence are also highly concerned about the training to be imparted to the prospective entrant. Thus it is surprising that during the last 40 to 50 years, the chemical engineering curriculum has retained a basic form and structure, which are accepted more or less over the entire world – something pointed out by Ruthven more than 10 years ago.

It is however true that, while the basic form has been retained, the details - courses, approach, methodology and emphasis have shifted considerably. And in the last two decades there has been sufficient growth of the field to merit some reconsideration. New areas like biotechnology, nanotechnology have emerged, demand for speciality chemicals and products is increasing rapidly, globalization has increased competition in business environment, communities have become more conscious about environment.

Boundaries of Chemical Engineering have been expanding in response to the advances in basic sciences, computational methods and technology and new societal demands like energy, environment, sustainability and in the last decade the expansion has accelerated. Changes in world political, economic and business situation and social awareness have also contributed directly or indirectly to these changes.

Chemical Engineers all over the world have certainly responded by conducting discussions, seminars, workshops and a host of recommendations have already been made- and some already are in the process of implementation.

In the US, the ABET system is primarily concerned with the quality of the training/ education imparted- and this does involve the content and curriculum- usually decided by experts, professionals and professional societies of the specific field. The accreditation method requires educational goals to be set for courses and the results/ degree of goal attainment evaluated. The method also emphasize skills in application and analysis.
The curricula and course content in the US is decided by the institute/ school – but it does follow a guideline evolving out of meetings, workshops, seminars participated by institutes, professionals and societies- like the AIChE-MIT workshop.

In Europe, the EFCE, based on the Bologna convention has made recommendations for the two cycle course- a 3 yr first degree and a 2 yr second degree course. A major difference is the duration- the first

degree- apparently bachelors- will require 3 yrs in contrast to 4 yrs in the US or India. But it appears that Bologna convention would make the completion of second cycle- that is Masters degree -the minimum qualification for a professional engineer. Thus it may be effectively going in for a 5 yr training.
The time scale is certainly important as many new topics/ subjects have been recommended for inclusion in the curriculum along with additional industry training.

In India, the IIChE had taken the initiative and had formulated a course/ curriculum guideline (CEEDC- Chemical Engineering Education Development Center) . The AICTE- the official accrediting body- has effectively adopted this guideline. There has also been several workshops/ seminars discussing the issue. Gandhi and Kumar in a recent paper have put forward a recommendation.

It is of interest but not surprising to note that the recommendations/ proposals of these workshop/ symposia are quite similar in broad content. They all generally agree that
i) the courses must be built around core chemical engineering , there being substantial agreement as to what the "core subjects" are.
ii) basic sciences- physics, chemistry , mathematical methods and multiscale analysis need to be emphasized, and some bioscience need to be introduced.
iii) emphasis must shift from " data/ information accumulation" (that is "memorize") to " analyze and apply", with stress on innovation and adaptation.
iv) sufficient flexibility to allow electives for introduction to different branches/ areas particularly the new and emerging ones should be incorporated.
v) there should be some courses for training in management, communication skills, enterpreneurship, social and environmental impact, sustainability
vi) industry apprenticeship and practical/ experimental work should be stressed/ increased
vii) there should be an evaluation of the quality of the education process itself

It is apparent that there has already been a great deal of discussion and a general convergence of opinion has been achieved, and now it is necessary to look at implementation rather than initiate further discussion. It appears therefore that there is little point in rediscussing the basic tenets- "absorb, assimilate, adapt and apply"- though what is to be included and what should be left out and how it is to be imparted will remain debatable forever. These aspects have also been addressed in many of the papers and presentations on many occasions including the present meeting(???) but still there seem to be a few aspects which have not yet attracted a detailed scrutiny.

Developments in Chemical Engineering

One of the major reasons for the proposals for curriculum modification is the rapid extension of boundaries of Chemical Engineering- and one aspect of this expansion is creation of new fields like energy, environment, biomedical technologies, atmospheric sciences. Now whenever an established field expands and sprouts out new areas, the field consolidates itself primarily in two different ways. The first one is broadly followed by fields of science, the second primarily by engineering. It does not however mean that all developments strictly follow only one model- it is just a rough classification.

Physics for example has expanded tremendously from its early days and particularly since Maxwell. A solid state physicist of today may be totally unaware of what is happening in particle physics and may not even need to interact. But all these diverse fields are still part of physics. This is reflected in the education system and course and curricula. At the lowest level, that is at bachelors level, there is more or less uniform training in the "basics" of physics. There are very few –if any- bachelors program in solid state physics or statistical mechanics. But in the next stage that is at masters level the specialization's are offered.

Compared to this, in many engineering fields- electrical engineering for example- when sizable expansions occur, new apparently "independent" fields appear which are generally considered to be separate disciplines. Electrical Engineering for example- when it started had no "electronics" component. Electronics was born inside it and when it became sufficiently "large"-it simply split out and is now considered a separate field. Similarly, computer science started as part of electronics and then became a discipline by itself. All these are now considered "independent" fields and separate bachelor level degrees are commonplace.

It is true that no field in its expansion rigidly follows one model, but a rough division is possible. Chemical engineering seems to emulate more of "science" model and all the recommendations for course and curriculum modification sited previously, favors this. While this is largely justified, splitting off some areas as " independent" subjects may be another way of coping with the recent explosive growth of chemical engineering. Some such splitting- particularly in areas which are "interdisciplinary" or "cross disciplinary" - have already occurred to some extent. Not only have Departments of Energy, or Environment have come up in many places, bachelors level programs have also been initiated. "Product" Engineering or "Molecular" Engineering may also come up. In the US, a PE examination in "Bioengineering Fields" is under active consideration. Such developments may end the monolithicity of Chemical Engineering , but may still be viable alternative and certainly will be explored.

Secondly the duration of training period also merits some consideration. The two cycle system in Europe may mean a 5 year training before joining the work force. A 5 year degree instead of the current 4 year should be given some serious consideration.

Globalization

Globalization does bring in a new dimension to Chemical Engineering profession - and certainly has strong implications on the training of chemical engineers. Globalization affects the chemical engineer in several ways- it makes them more mobile- being trained in one country and serving in another. They may also get part of the training at one place and rest at another. This requires a certain amount of uniformity in the training that is uniformity of courses and curriculum. And it also requires certain amount of uniformity in nomenclature- so that the course content is properly reflected in the name- and also some uniformity in the extent and depth of coverage.
Secondly it becomes necessary for the student to get acquainted with the laws, customs and practices of different countries. Of course the incumbent engineer does not need to learn the practices of every nation, but there should be some training that will make him able to learn and adapt to the situation.
Communication skill will be another aspect of training that is strongly affected by globalization.
Thirdly, the social and environmental impacts will require greater emphasis during training. Local and global effects of chemicals and chemical plants and social interaction/ effects will need to be carefully evaluated.

Thus the need for certain amount of uniformity is recognized by almost all schools all over the world, and some significant regional efforts like the Bologna convention- have been made, but there has not been any similar effort on global scale and lead in this matter need to be taken up by the well established professional societies.

While Globalization generally implies some sort of uniformity across the world, it may also harbor diversity- depending on the concentration of expertise at different locations/ institutes. Since the services will be available on a global basis, specialization may also flourish to much greater extent.

Harmony with environment and society

Possibly the most important lesson that Globalization teaches us is that we cannot live in isolation. Every individual, group, country or region is connected to every other in a very intricate manner and every action we take affects the rest of the world in some way.
It is now felt and accepted that "Chemical Engineering" and " chemical Industry" has to be in harmony with the rest of human existence and the human ecosystem. It cannot remain a separate compartmentalized field of activity disjointed and isolated from rest of human ecosystem.

A broader view of professional ethics would require the Chemical Engineer to be broadly responsible for all the consequences of his action, both individual and collective, and be responsive to the rest of the society and the entire ecosystem.

Globalization, making information flow very fast and effective, has also made us aware of the great diversity of view points and interpretations that exist in the world. Many of these world views may even be directly contradictory, yet each have its place in the tapestry of human domain. Learning to accept the existence of different world views is now necessary and the ability to understand and appreciate the "other view point" is desirable.

Ethics courses- many of which are in existence- and are mainly catered to by non-engineering Departments, may not truly project the ethical issues and conflicts a chemical engineer has to face in the context of operation of a plant or organization dealing with chemicals. But survival of the world now demands that we match our "wants" with our "needs". . While this has to be decided by the entire world society, we have to realize that "entire society" does include "chemical engineers", "chemicals" and "chemical plants" as well. So the Chemical Engineer has to take part, and use all the information and apply all the knowledge he/she has. But where or how does it fit in the context of "Chemical Engineering Education".

Teaching the concept of harmony and imparting a motivation to achieve it certainly cannot be classified as "technical part" of the training. But in order to practice it, a host of technical skills and knowledge becomes necessary. The chemical engineer needs to take a critical look at various processes and phenomena that does come under his field of knowledge and make this knowledge and analysis available to global scientific community . Thus he has to take a critical look at the chemical industry, and also chemical processes occurring in the environment, feasibility's of products, processes and their by-effects. It is this aspect which requires training and hence becomes relevant in the present context- that is "Chemical Engineering Education". .

Thus at the base level certain technical training becomes necessary or at least desirable. The engineer should know the effects - both long term and short term- of the products and the process plant on the human ecosystem. Long term effects need life cycle studies of the products, all possible effluents from the plant, effects of raw material procurement and problems like transport losses, accidents. That is the engineer must have a holistic view even at the conceptual design stage. The engineer must be trained to think naturally of costs to include these effects as well .

Thus it requires a change in attitude and also knowledge and capability for computing costs of all these factors. Normally courses on environment, pollution or waste treatment do not include this attitude or training. But these should be included at some stage of the training, and the methodology should be made an integral part of design course.

It must be realized that life cycle studies or associating costs with such effects are not the easiest of things to do. So possibly the best one can do at present is to inculcate the attitude- incorporating it as an integral part of curriculum. Quantification of such items require application of basic chemical engineering and chemical and biochemical principles- namely reactions, kinetics of reactions, heat and mass transfer, flows and mixing. So courses in heat transfer, mass transfer, modeling etc. should contain sufficient examples of applications of these basic principles to problems from these fields with as much real data as possible. Evaluation of cost of product including the cost to the society, environment and ecosystem is also a difficult task and requires training. Economics and management courses should therefore contain discussions and methodologies for such evaluation- at least the basic concepts and methodologies. Apart from economics, such cost evaluation would involve considerable amount of statistics and associated mathematics- and elementary concepts need to be introduced in mathematics and statistics courses.

An integrated curriculum is required to enable the student to have a holistic view and be able to see the interconnections between the different courses and fields- and instead of seeing each part as a separate entity, he should see a "continuous whole"- and single field. And this field should blend "seamlessly" with "life", "reality" and human ecosystem.

Two other aspects also seem important. Firstly, there should be some awareness about " alternate materials" and "processes". This may be introduced as an elective course- and this course should at least include qualitative ideas about alternate materials and processes.

The second aspect is communication. This is already recognized and well stressed and included as part of curriculum in many programs. But in most of such courses, the emphasis is for communication with peers that is people in engineering, administration and management or at most with "labor" or "non technical personnel" serving chemical plants. But the skill to communicate with a non-peer- that is an ordinary person, who is not a part of "chemical industry community" or related to it, and has no or little knowledge about these, is also important. The engineer should be able to communicate with such persons, understand his concern and appraise him of the technicalities, in a way that he "understands" and finds correct answers to his queries. Thus the engineer also needs to have some competence in "popularizing science and technology"- so that he is able to communicate the facts in an undistorted manner to the uninitiated. And in doing so- this is where ethics do come in- he must be strictly honest, impartial and unbiased.

Some Suggestions

Needs for courses which deal with multiscale behaviour, process enhancement, analysis of chemical processes , product design with desired end properties, linking science with engineering practice with focus on modelling skills, creativity and ability to approach and solve poorly defined problems are well recognized and are being implemented, but social responsibilities of the chemical engineer must also be integrated into these courses.

Many departments and programs of courses do include instructions on environment, alternate technologies, nonconventional energy, - but usually in form of a separate courses. That may not be adequate- partly because most such courses are electives and there is no integration with the core chemical engineering courses. Such integration possibly requires that problems and student exercises for core courses be chosen from these fields. It also needs some changes in the simulation, modeling, design and project engineering courses. While most the basics or fundamental principles or methods may remain the same, the scope or the application field of these courses need to be enlarged.

At least one important new field need to be introduced- that is the methodology of valuation or ability to attach numbers to "fuzzy" items about which only "qualitative" knowledge is available.

In conclusion globalization does have direct and indirect effect on how the professionals should be trained. The increasing liability and the wider arena of activity requires apart from training in adaptability, ability to design and produce products meeting the exact requirements in quality, quantity and variety to satisfy the society.

In addition, there has to be a change of attitude- chemical engineering and chemical industries should not be looked upon as an independent self standing and separate institution but as an integral and harmonious part of the entire world society and human ecosystem.

"Chemical Engineering Education- a personal view", Chem Engg Sci 51, 18 iii-iv (1996)
"Athena, Hercules and Nausica: Three dimensions of chemical engineering in the twenty-first century", Fluid Phase Equl 261,3-17 (2006)

212c Chemical Engineering Curriculum for Sustainable Development Vision 2020 - Design & Engineering of Plant & Equipment

K. Venkataramanan, Larsen & Toubro Limited, Mumbai, India

The present day chemical engineering curriculum evolved over the last century incorporating concepts like unit operations, process control, kinetics & reactor design, and transport phenomenon. Later landmark technological events like biotechnology, personal computing, microelectronics, advanced computing nanotechnology are driving chemical engineering from macroscopic towards microscopic to nanoscale and molecular dimensions. While fundamental definition of chemical engineering has remained same, its scope has expanded significantly.

Chemical engineering scenario is changing rapidly in 21st century. Chemical companies today are highly diversified and knowledge based relying on technology, science and innovation to create value added products. Some of the basic sectors like petroleum refining, power plant engineering, and cement are witnessing large capacity installations. Refinery of 21st century apart from using novel concepts like reactive distillation and membrane separation will also look at biotechnology for bringing improvement in chemical conversion and efficiency. New developments for efficiency improvement in power plant technology like combined cycle / once through boiler / supercritical technology and IGCC will need successful development of advanced materials for higher duty. Chemical Engineering research will need to be interdisciplinary integrating knowledge from traditional chemical sciences, biotechnology, and material technology. This will need curriculum to be more interdisciplinary incorporating the disciplines of biology, biochemistry, physics and computational methods.

Process and product development, starting from a basic idea, steps through research, laboratory, bench / pilot scale and scale-up to final process. Accurate models based on sound theory with process modeling tools significantly reduce scale-up and marketing time and cost of new products/ processes. This is also the current trend for meeting challenges of reactor design in complex plants of ever growing size, though it still needs integration with metallurgy selection, advanced manufacturing methods and high end stress analysis. Design softwares and codes like HTRI, HTFS and CFD has enabled optimizing heat exchanger sizes and material selection.

Plant design includes all engineering aspects involved in the development of commercial process in a chemical plant. Involvement of chemical engineer starts from development of basic engineering package or carrying out front-end engineering and design activities. This FEED activity covers activities like design basis formulations, definition of appropriate codes and standards, process simulation, detailed PFD and P&ID preparation, material selection, equipment and line sizing. Chemical engineer also integrates inputs from different engineering disciplines like mechanical, structural, electrical & instrumentation and piping. Chemical engineers need to be acquainted with the general problems and approaches in each phase of plant design. Plant design also considers aspects like safety, health and environment concerns. Chemical engineers must be versed with statutory norms involving environmental protection, workers safety and health.

Advances in computational technologies are improving the application of fundamental sciences. The most critical for the chemical industry are process modeling, simulation, operations, optimizations and control. Static and dynamic modeling/simulation along with optimizers achieves optimal design and operation. Use of computer softwares has become indispensible part of process design. Advances in computer hardware / software coupled with fundamental knowledge will change the way processes are designed and plants are operated. Faster incorporation of computer based ideas into curriculum will enable its use from design phase to commercialization.

The rapid development of process industries has given rise to multiple environmental issues. Incorporation of sustainability concept into industrial operation has become important to the future of corporate performance and for public acceptance of industrial facilities. Chemical engineers are in a position to make significant contribution to achieve sustainable development by promoting sustainable chemical plant design, eliminating the need of toxic chemical usage, reducing waste and improving process operations. Chemical engineering tools like steady state and dynamic simulation, advance process control, pinch analysis helps in design of more sustainable product and process. Sustainability concepts/tools like waste audit and inventories, health and environmental impact assessment, life cycle assessment, sustainability matrix help in evaluating sustainability of a technology and benchmarking sustainability index of operation. Presently sustainability is treated as add-on concept and not integrated or embedded in the core subjects. Embedding sustainability and sustainability evaluation tools with relevant chemical engineering subjects will enable to develop sustainable products/process in design stage itself.

Skills and knowledge which chemical engineer needs to acquire as a part of curriculum could be such that it aligns with technology vision/ needs specific to the country or the region. India with its large population base needs to concentrate on sectors like food, water, energy pharmaceuticals and minerals. The purpose is to ensure sufficient availability of basic needs at affordable costs. Chemical engineers have a pivotal role to play in developing and applying appropriate technologies and develop products using locally available raw materials. Industry and academia need to work seamlessly in identifying and executing the projects, which will benefit the masses.

Chemical engineering graduates are facing a new environment as a consequence of technological developments, globalization and widened scope of modern chemical engineering. Challenges before chemical engineers are to develop products / process and produce energy in a sustainable manner. Curriculum developed should therefore provide engineers with balanced education in technical as well as non-technical disciplines of engineering to broaden their skills and knowledge that will satisfy the needs of industry and society.

Role of Public Funding In R &D

Rajiv K. Tayal

Department of Science & Technology, New Mehrauli Road, New Delhi – 110016, India

e-mail: tayal@nic.in

Abstract

The level of expenditure incurred on research and development plays a significant role on the overall health and long term sustainability of the economy of a nation. Public funding plays an important role in supporting and sustaining research and development. The role is more direct and feasible in case of basic research carried out primarily in academic institutions. Nevertheless, the impact of public funding can also be established through an indirect correlation with the research carried out in research laboratories and in industries. Similarly, the role of public funding could be more critical in case of R & D activities in a developing economy than in a matured economy.

1. Introduction

A high percentage of public spending on R & D goes to support work at the grass-root level. Hence, it is difficult to establish a direct and objective correlation with the output. Most of the open ended basic research in academic institutions is supported with public funding. This work results in intellectual output which either goes in public domain by way of published papers or is protected through IPRs. The very nature of this output is devoid of physical attributes like product or technology and hence its visibility remains rather diminished, particularly beyond the scientific community. Fortunately, the importance of public funding on R & D is amply appreciated in abstract sense and serious questions to justify this expenditure are seldom raised. Nevertheless, any reasonable attempt to correlate the input which goes into R & D and the output which fructify always makes an interesting reading. The present paper makes an attempt to provide an insight into the issues involved in the entire value chain of research & development and thereby trying to establish the contributions of public funding, for which an extended correlation is necessary. It is realized that complete objectivity may still remain an illusion, however, trying to see things at the micro level would facilitate a better appreciation of the overall perspective.

2. Classification

For the purpose of this analysis, broad term R& D has been classified as follows.

- Basic Research
  - Open ended
  - Reverse Engineered
- Applied Research
  - Routine Problems
  - Exploratory
  - Comprehensive Technology Development

It may be noted that there is nothing sacrosanct about the above classification. The purpose of this classification is neither to compartmentalize research nor to standardize the nomenclature. It is only for the sake of convenience that this classification is attempted so as to examine the issues involved in each one them and establish the role of public funding there- in. Their collective view would in-turn provide a better overall perspective to appreciate the role and impact of public funding over the entire landscape of research & development.

It is also important to see that a fully water tight compartmentalization is neither possible nor worthwhile in this case. The first category of 'basic research' is mostly concerned with the 'R' (Research) portion of R&D. The second and third components of 'applied research' are predominantly concerned with 'D' (Development) in the ascending order. Many a times, we see a mix of both 'R' & 'D' in such cases. Nevertheless, the balance is normally skewed towards 'D', in the later two cases at least in terms of physical effort. In a typical extreme case of technology development, it assumes the dimension of purely engineering challenge, needing enormous effort and integrated approach.

However, it is quite obvious to visualize and common to see basic research as a feeder line to applied research and technology development. The findings and outcome of basic research become obvious starting point for process development and scaling-up. Similarly, the basic research is quite often inspired by the needs of applied research and product & technology development. In that sense, strong interfaces do exist between the categories and their complimentary nature is fairly evident.

We may now attempt to analyze each individual element of the above classification to examine their respective attributes, issues involved and the role of public funding therein. The same analysis is also presented in summarized form in Table -1 underneath.

Table- I: Research & Development Matrix

Category	Type	Environment	Motivation	Impact	Support
Basic Research	Open Ended	Academic	Natural Curiosity	Low-Moderate-High	Public
	Reverse Engineered	Academic Research Labs Industry	End Application	Moderate- High	Public Public – Private Private
Applied Research	Routine	Industry Research Labs Academic	Immediate Problem Solving	Low- Moderate	Private
	Exploratory	Industry Research Labs Academic	Improved Product, Process or Practice	Moderate- High	Private Private- Public Public
	Technology Development	Industry Research Labs (Academic – specific top up role)	New/ Improved Technology	Significant	Private Private- Public Public

3. Basic Research

Open ended basic research is purely guided by the natural curiosity of man into the mysteries of nature and natural phenomenon. This may or may not lead to some meaningful conclusion, particularly in terms of material utility or commercial value. Even in the case of success, the short term application potential or commercial value can not be predicted. However, there is always a possibility that the basic findings may lead to a vast number of applications, although over a long time horizon. The immediate contribution is of course in terms of enriching the existing knowledge base in the concerned area. This kind of research is normally seen in academic institutions and remains entirely in the domain of public funding. Research laboratories may contribute in a marginal way depending upon their institutional priorities and availability of resources for exploratory work.

Reverse engineered basic research is guided by some end application in mind and that determines the issues to be investigated and resolved. This type of research could be prevailing in academic institutions, research laboratories and industry. In fact, there is a good reason for industry to take initiative into the research of this kind because this may lead to major breakthroughs in terms of process, technology and product development. However, this type of research is likely to be taken up only by those industries which are market leaders in their own segments and have some sort of core competence in that area. The following are obvious reasons.

- They have strong motivation to undertake basic research, achieve significant breakthroughs and stay ahead from competitors & followers to retain their position of leadership.
- They have adequate resources and the risk capacity to right – off investment in case of failure.
- They also have adequate resources and critical mass to exploit the full potential of the discovery and reap the windfall gains, it may accrue.

There is no guarantee that each such endeavor will yield expected results. However, an optimal mechanism is worthwhile to institutionalize to capture the possibility of reasonable number of successes with the help of a carefully designed –"organized research programme". The other dimension is of the possibility of few spin-offs of commercial value towards an overall justification for investments in such a mechanism.

4. Applied Research

Applied Research corresponds to the use and application of established knowledge to demonstrate a new process, product or technology or optimize/improve an existing process, product or technology. The first element of applied research is concerned with day to day problems of routine nature pertaining to a specific industry. One can also use a less sophisticated term "trouble shooting" for this activity. These problems do not justify deployment of public funds and they ought to be handled by the industry itself.

The second category of applied research could be of more challenging and exploratory problems, involving altogether new or significantly different approach or methodology to develop a process or product. Since the commercial angle is evident in such cases, the primary objective of quick results and their rapid deployment into commercial practice most often overshadows the entire plan. Time and speed of delivery rather become the benchmark. That leaves little scope for extensive investigations into at least few basic issues with the aim to impart substantial value into the entire effort. However, this approach may be self-limiting in the holistic sense. Trying to find quick fix solutions using engineering intuition may be alright in short term but does not complete the entire research loop in its logical conclusion. Deeper insight into the problem may not always yield substantial advantage in commercial term. However, it is good to institutionalize such practice for the overall health of a research outfit.

Most of the private funding in industry on research and development is directed towards the two aforesaid categories. Private funding goes into this kind of research either by way of in house R& D or through sponsored projects to academic & research institution. The mode of execution will depend upon the availability of in house expertise and the overall priorities. Normally, an industry would go for research outsourcing for any one or all the following reasons.

- For the want of highest end intellectual inputs at the most fundamental level.
- Work/activity is relatively trivial as compared to their overall scale of operations.
- An intense exploratory effort in a new area or of altogether different kind may by a serious distraction to their core activities.

Public funding is available for this activity as well, if it is has enough substance to make significant impact at the sectoral level. There could be a strong case for public- private partnership in this category, depending upon the kind of challenges and risk involved and the significance of the outcome in terms of the entire industry sector and the economy. Research laboratories participate in this activity in a significant manner, if it comes under their area of core mandate. Support for this may come either by way of public funding or industry sponsorship.

There could be a third dimension to applied research in terms of a comprehensive effort to develop a cutting edge technology towards a new product or significant improvement in properties/ attributes of existing product or cost of production. This kind of effort would naturally involve significant deployment of resources and an integrated coherent team approach, involving interdisciplinary expertise from different functions. Industry is ideally suited and organized for this activity, may be with the help of academic institutes and research laboratories for specific inputs and expertise. Leadership and managerial resources assume equal importance for successful realization of such endeavours, besides the technical expertise. This kind of activity is again confined to the market leaders in their respective areas for the same reasons as explained before in case of reverse engineered basic research. We do see a parallel activity along these lines in case of strategic areas like space, atomic energy and defence where public funding is exclusively used to support research and development. Nevertheless, the character of approach for the execution of such tasks remain the same.

5. Role of Public Funding

In the foregoing text, we have seen the role of public funding in various facets of research and development. While the open ended basic research is exclusively supported by public funding, major portion of application oriented basic research is also supported by public funding. As an extension, the influence of public funding can be indirectly established in applied research, since it is based on the foundation of sound basic research.

Public funding assumes the role of an arterial lifeline to create and sustain a live environment for research and development in academic and research institutions. The manpower which is eventually deployed by industry for research and development is trained with the use of public funding in these institutions. The skills acquired by students in higher end academic programmes (post graduation, doctoral and post doctoral) in handling sophisticated instruments, designing & conducting advanced experiments, proper analysis & interpretation of experimental data to gather meaningful conclusions and modeling & simulation would make them suitably qualified & trained to take up research and development in the corporate sector. Public funding is also deployed to initiate work in areas which are exploratory and involve high degree of risk, till basic feasibility is established, market sensitized and industries are willing to take it forward.

The role of public funding is more critical in developing economies to sustain research and development. In developed economies, the industry is matured and there are several market leaders in each segment to take a bigger initiative on their own, not only in applied research but even in basic research, to counter competition and retain their position of leadership.

6. Conclusion

We have come a long way since the time when R & D support was near exclusively the domain of public funding. Today, the R & D expenditure in industry is expanding both in absolute terms as well as a percentage of their top-line & bottom-line. In the present world order, when more and more nations are adopting the model of market economy, the industry is bound to be a more active partner in the overall scheme of things. Nevertheless, the role of public funding will continue to remain important for supporting basic research and in high risk or strategic areas. We may also see greater number of public-private-partnerships for supporting R & D activities in the times to come.

BACK TO BASICS

Dr. S Ganeshan, Toyo Engineering India Ltd, Mumbai, India
and
President, Indian Institute of Chemical Engineers

Introduction

In the 1967 movie *The Graduate,* Benjamin Braddock is advised to take up a career in plastics: *"That's the future, son."* Plastics, after all, defines us and defines the modern world. But, Chemical Engineering as we know it will in all probability cease to exist in the near future, though this does not mean the end of the discipline. Chemical Engineering has morphed and mutated but basically it has remained the same. This might seem like a contradiction but it is true nonetheless. It is the endeavor of the author of this paper to present his views as one who has been involved in almost all aspects of the practice of chemical engineering. Perhaps the credentials of the author are pertinent at this stage.

I have been heading the process design department of an engineering company that serves the Chemical. Petrochemical, Fertilizer and Refinery sectors in India. I have been involved in recruiting Chemical Engineers, both fresh from university and also experienced engineers. I also train new hires. In addition, for almost ten years I have also been an Adjunct Professor at the Indian institute of Technology, Bombay, where I have been teaching core chemical engineering courses, both independently and in association with other faculty. I do this during my normal work hours, which is a form of commitment by my organization to the development of the profession. To top it all, I present this paper in my capacity of President of the Indian Institute of Chemical Engineers. Thus, I have had a hand in making, customizing and utilizing the product called the Chemical Engineer, and presumptuous though it may seem, I feel I have an interesting tale to tell. .

Historical background

Chemical engineering education in India has almost entirely been modeled on the American pattern. For historical reasons, most branches of engineering in India have been based on the British pattern. The exception has been Chemical Engineering education, which in its current form is relatively new in India, being about 60 to 70 years old. Perhaps in the flush of political independence following the end of British colonial rule in India, we tried to look at the other option available to English speaking engineers: the American model. Thus, all Indian Chemical Engineers have cut their professional teeth on American textbooks, starting with Unit Operations by McCabe and Smith, Mass Transfer by Treybal, Transport Phenomena by Bird et. al, and of course, Perry's Handbook, etc. These worthy books and other American texts still form the bedrock of Chemical Engineering education in India. A point to note is that this trend continues though very good books — notably the Chemical Engineering Series of the British authors Coulson and Richardson — are freely available. Still, we prefer the American pedagogy. This is true on a pan-India basis. Thus, the Indian

Chemical Engineering curriculum and the issues facing it mirror the problems of American academia.

The nature of the curriculum of Indian Chemical Engineering

Currently, the undergraduate engineering program consists of eight semesters of 16 weeks each, and is spread over four years. Entry to the program is after 12 years of school. The entrants have a grounding in Mathematics, Physics and Chemistry, and an introduction to calculus. The first-year engineering student is not less than 17 years of age. On an average, each semester is comprised of five courses excluding laboratory work. The first two years are predominantly devoted to Mathematics, Physics, Chemistry, language, social sciences, etc. This leaves only about 20 courses in Engineering, a significant part of which will necessarily have to be general engineering (Electrical, Mechanical, etc). There is no alternative to this: if we still cherish the label of 'Engineers' then there is no escaping the basic engineering tools viz., Mathematics, Physics, Applied Mechanics, Strength of Materials and Electrical Engineering. I would like to insert a caveat: this is a broad description and there is a significant difference in how various institutions conduct their courses. The university system differs markedly from that of the autonomous Institutes of technology. There is a wide variation in the course content and pedagogical methods across institutions.

Chemical Engineering by its very nature needs a strong grounding in general subjects: more than other disciplines of engineering it is a truly interdisciplinary engineering course. So, willy-nilly, the chemical engineer has to spend a lot of time on 'non-core' subjects. The net result is that this leaves very little time for any specialization and today, specialization is essential.

The Environment

Speaking from the Indian perspective, Indian Chemical Engineering is only now coming of age. Even today, almost if not all industrial technologies are either European or American in origin, with American technologies being predominant. Indian engineering effort is still at the tertiary level with basic engineering and know-how coming from abroad. This is true of all large capacity capital-intensive production facilities. Despite India having a very large Chemicals sector, there has been very little world class technology developed in India. In India, what we are doing is more in the nature of residual and detailed engineering, which requires more tertiary and volume-intensive skills. But we perceive a change: witness the pharmaceutical sector, where from copycat technology one is beginning to see more of research and development at the molecular level. A new business stream is contract research with many big names setting up elaborate establishments in India. India is now a favored destination for such activity. The reasons are pretty clear. Low costs are only part of the story. The availability of technical personnel in fairly large numbers was a plus point but protection of Intellectual Property Rights (IPR) was a major issue. Once India became a signatory to the WTO patent regime which protects IPR on both product and process know-how, it opened the floodgates for contract research and also gave the necessary reassurance to technology leaders enabling them to be more open with their know-how. That India is a democracy where the due process of law prevails, is an important factor. All this bodes well for the fresh engineer provided he* has the wherewithal to exploit the situation. The rub is how to provide him with the tools for the trade.

The student

As mentioned earlier, Chemical Engineering is a long haul and requires an extended amount of study to lay the foundation for building a proper superstructure. In this respect, chemical engineering differs from other branches like mechanical, civil or electrical engineering where the undergraduates are good enough for direct employment, at least in not very technically intensive fields. The uniqueness of Chemical Engineering is that unlike the other branches of engineering that are all derived exclusively from Physics, it is the only discipline that also leverages the unique powers of chemistry to solve 'Engineering Problems'. Chemical Engineering is the only discipline that manipulates material at the molecular and atomic levels. Hence, a student of chemical engineering needs to spend a lot more time building the basics, which will include large dollops of chemistry. Of all engineering streams, chemical engineering has the largest areas of commonality with "new" sciences like Biotechnology and Nanotechnology. Chemical engineers are best positioned to exploit these sciences, for the betterment of life as we know it.

Chemistry

Historically, chemical engineering grew out of chemistry. The original chemical engineers were necessarily chemists with good business acumen. After all, engineering is making a product at the least possible cost. It worked fine so long as the scale of operations was small and processing, batch-wise. Continuous processing and the understanding that increase in scale brought its own economies meant that chemistry retreated into the background in chemical engineering operations. The commercial scale of even the immediate post-WW II era was little more than a pilot plant of today in orders of magnitude. The 'engineering' problems of the older generation plants could be resolved by a reasonably smart industrial chemist. But let us not forget that this practitioner had a strong grounding in chemistry. With the scale of operations increasing, the chemistry of the operations became less important and chemical engineering as defined by unit operations gained prominence. Hence the well worn adage that a chemical engineer need not know chemistry. A lot many chemical engineers were perversely proud of not being good at chemistry. The wheel has now turned full circle. Reaction mechanisms, bio-reactors, enzyme catalysis and process intensification are the new buzz words, and once again chemistry has come to occupy the high ground in the chemical engineers' world. So, to come back, the chemical engineer of today and tomorrow needs to study a lot more chemistry than the practicing engineer of yesteryears. Even the nature of the chemistry required has changed and we need more of biochemistry, molecular biology and thermodynamics. The problem is where to squeeze this requirement in the already crowded curriculum.

The users' perspective

The perspective of the layman, and also that of any prospective employer of a chemical engineer, is that he is an engineer first and so is expected to have a decent grasp of mechanical, electrical and civil engineering principles. Not surprisingly, no one expects a civil/mechanical/electrical engineer to have even a modicum of understanding of chemical engineering. Thus a chemical engineer needs to spend a lot more time learning the basic things: Mathematics, Physics, Computer science and of course, a lot of Chemistry. He also needs Applied Mechanics, Strength of Materials, Electrical Engineering and an understanding of Theory of Machines. Like all engineers he needs an understanding of Economics, Management and Operations Research. Furthermore, no engineering education can be complete without an exposure to the liberal arts, sociology, etc., because, ultimately we need to prepare good citizens as much as we need to prepare good chemical engineers. These are the basic building blocks of chemical engineering and no matter what else we teach them, the basic core has to be forged. It is only after all this is done can we even think of chemical engineering or as I want to put it, basic chemical engineering.

Basic chemical engineering

Basic chemical engineering must cover transfer processes (fluid mechanics, heat and mass transfer), thermodynamics, reaction engineering, process control, unit operations and equipment design. Without these units one fails to see how a graduate can be considered a chemical engineer. Let us not forget that this is the minimum requirement. But the contention is that in today's milieu, such an engineer will be fit for only the simplest of jobs at the very bottom of the value chain, as the requirements of the industry have become extremely complex. In the era of flexible work environments, every engineer needs to be a multi-skilled multi-tasker. At the same time the Chemical Engineer must also specialize if he is not to stagnate at the bottom of the pyramid. It is only after this rather heavy meal of basic chemical engineering can we even think of specialization

Product mismatch

The fresh chemical engineers that I have interviewed are all too often familiar with stochastic differential equations and fuzzy logic theory but do not understand pump NPSH; they can manipulate a process simulation package but have no understanding of the Gibbs phase rule. They can work around HTRI but do not understand LMTD. In short, what we see is a lot of information but not much scientific knowledge or the ability to adopt and adapt knowledge at hand. Time and again, we get engineers who cannot 'think on their feet'. I repeat, we have a good looking superstructure built on very shaky foundations.

In most chemical engineering curricula in India, we have a 'Home paper' where a student is expected to design a plant for a specified chemical. This calls for an extensive survey of the literature, flow sheeting and design of major equipment, topped off with a financial analysis of the proposed project. Now, one finds almost all the information in the dissertation to be sourced from the internet without any assessment of the reliability of the information provided.

Take the case of trawling for information. Those of my vintage know the pain and pleasure of combing through the 'Chemical Abstracts' for information. Though a mouse click on the net is much faster, is it a better alternative? I do not decry these new tools, but there is no substitute to understanding the fundamentals.

Is there a better way? My contention is that what we need to do is spend more time on the basics. Chemical Engineering is surely changing but some things remain the same. Today the stress is more on the new engineering which is not a bad thing in itself, but unfortunately it is coming at the cost of basic knowledge and this is the pitfall that we need to avoid. We need to stress on basic engineering and of course on Chemistry. Ultimately, it is Chemistry that defines us. But the chemistry we need to impart to NextGen is different. This Chemistry should be biased towards Thermodynamics, Surface and colloidal Chemistry, Electrochemistry, Biochemistry, Catalysis and Materials Science. Specialization at the undergraduate level is counter-productive as the Chemical Engineer needs to have a reasonable depth in other disciplines of engineering also. Unfortunately, what one sees is a reduction in emphasis on the core courses mentioned above, in favor of the newer and non-core subjects. Received wisdom suggests that such courses would attract more students. Whether the premise is true or not, the fact is that the end product is virtually unemployable in the conventional industry which still needs conventional nuts and bolts ChEs. Not surprisingly, the student chemical engineer is a very confused person. He is nearly half way through the four year program before he develops even the vaguest of ideas of what chemical engineering is all about. Simultaneously, the student is presented with a wide variety of unrelated electives and so behaves just like a child in a sweet shop: he wants everything. He is fed so much of other 'soft' information that he has all but lost the appetite for rigorous technical analysis. By now the initial enthusiasm is lost and the student is ripe for poachers (the IT sector being the biggest attraction). What we need to do is: motivate the engineer through the basics; explain how Chemical Engineering is 'different' from other branches of engineering; get them used to the inevitability of a specialization after the basic degree. Most universities in India got it right when they considered Chemical Engineering distinct from other branches. Now the tendency is towards amalgamation and to label chemical engineering as just another branch of engineering. This is wrong. To repeat, Chemical Engineering is different.

Specialization

As mentioned earlier, the future of chemical engineering is in specialization but this specialization can happen only after the undergraduate program is complete. This is similar to the model adopted in medical schools where virtually no doctor voluntarily terminates his academic pursuit after the basic degree. The medical student knows very early that specialization is necessary and is also made aware that this can happen only after the basics are mastered. Medical specialization is not done at the undergraduate level as there is simply too much to do in general medicine and this leaves no time for specialization.

Some initiatives

Earlier, plant visits used to be a high point in the engineering curriculum. Today, for various reasons entry to most plants is restricted. Hence it is necessary for organizations not subject to such concerns to take up the slack and encourage strong interactions with the academic world. Industry-academia exchanges in the form of lectures need to be stepped up to enthuse and encourage the aspiring chemical engineer. IIChE is also doing its bit by reaching out to the students in various ways. A recent initiative is having a separate annual convention only for students, which is highly popular. We need more student specific activities, if only to make them feel that they belong to a highly sought-after fraternity..

Conclusion

To sum up, Chemistry defines us. It is our claim to exclusivity. It is why we are different. It is why we are. We must ensure a firm foundation in the basics at the undergraduate level and leave specialization to the graduate degree. We must say to the young chemical engineer, *"Go back to the basics, son."*

*Note: The use of the word 'he' is not intended to be gender specific

212f Chemical Engineering Education in India - a Comparative Study with US

Jayant Modak, Department of Chemical Engineering, Indian Institute of Science, Bangalore, India

In India, engineering is one of the preferred choices for good students ready to undertake the undergraduate education after 12 years high school/junior college education. Rapid economic growth (8-9%) in past several years has fuelled a demand for engineering graduates. The enrolment of undergraduates in chemical engineering in India has steadily increased and currently about 150 colleges produce 8000 graduates year. In globalized market place, the engineering education should not only address the local demands but also prepare graduates to compete internationally for gainful employment. There is a considerable debate and discussion in chemical engineering community world-wide about the challenges faced in the globalized market place and the need to reform the chemical engineering education. Indeed, several concrete steps such as redrafting the curriculum with focus on molecular sciences, new teaching methodologies and their assessment the effectiveness of teaching, and imparting self-learning skills to the students have been implemented in several countries. This presentation will address these issues as they relate in particular to India. Government funded institutes and programs are growing in number but the rate of increase is not commensurate with the demand for enrolment in engineering programs. As a result, the participation of private sector in undergraduate education has increased several folds. The growth in the number and size of such programs has led to the growing demand for appropriate engineering educators. This has also led to a certain degree of heterogeneity in quality of the chemical engineering education in India.

This presentation analyzes the current trends in chemical engineering education in India and discusses some strategies that needs to be adopted, and compares several aspects with the situation in the US, which is quite different, and where well established Departments and Institutes are accommodating the changes in various ways.

258a Evolution of Chemical Reaction Engineering

Octave Levenspiel, Oregon State University, 102 Gleeson Hall, Corvallis, OR 97331-2409

Until recent years knowledge of chemical processing was descriptive and qualitative.

In 1810 modern chemical theory was born and process description became quantitative.

Then about 1900 the quantitative engineering approach was developed, first for physical changes, called the Unit Operations, and somewhat later for chemical operations. This we call the American approach.

In 1957 European chemical engineers brought together the design of chemical and their related physical operations under the name of Chemical Reaction Engineering, or CRE. This approach and name received practically universal acceptance.

Today the methods of CRE are widely used in the processing of biochemical and all sorts of other systems,

This talk wanders through this development.

Chemical Reaction Engineering (CRE) Education: From the Era of Slide Rule to the Digital Age

H. Scott Fogler, University of Michigan
Michael B. Cutlip, University of Connecticut

Abstract

The complexity level of CRE problems that students work and study in both undergraduate and graduate courses has increased very dramatically in the last 40 years. In the era of slides rules, the major focus was on reactors operated isothermally where charts, nomographs and linear differential equations were used to describe very simple situations. However, today's students are routinely solving complex problems involving multiple reactions with heat effects in CSTRs and PFRs that are either operated adiabatically or have co-current or countercurrent heat exchangers. The ability to solve such complex high-level problems has been enabled by user-friendly software packages, such as POLYMATH, that easily allow solution of systems of ODEs. Numerical solutions of CRE problems can easily be incorporated into example problems and special simulation where many "what if" scenarios can be examined very quickly. These provide for not only a greater understanding of CRE but also serve as a vehicle for students to develop and practice their creative thinking skills.

The application of CRE fundamentals has been extended to many new and emerging areas such as biological systems (fermentation and enzymatic reactors), electrochemical systems (fuel cells), combustion processes, pharmacokinetics, and environmental engineering, to mention only a few. As new technologies emerge, it is clear that the fundamentals of CRE will continue to be applied and will make enabling and substantial contributions.

1. Typical Course Content in CRE

Chemical Reaction Engineering (CRE) is a core course in most every chemical engineering department in the U.S. and perhaps the world. CRE along with Separation Processes are the two primary courses that help to distinguish chemical engineering from other engineering disciplines. The evolution of CRE to its present level of importance and popularity started in the 1940s when CRE was barely mentioned in any chemical engineering courses.

In the early days, CRE was known and referred to as chemical kinetics and catalysis. As the number of different chemical products steadily increased, so did the production rates of these chemicals. It became apparent that the equipment and processing that worked for making one chemical did not come close to meeting the demands of making a different chemical. Everyone knew they could not just use the same type reactor feed, and temperature for a different reaction but very few chemical engineers had the CRE knowledge to scale the reactions. Pilot plants were used to determine the processing conditions for the full-scale production of the desired product. The experiments carried out on the pilot plants to find the right temperatures, processing times were costly and time consuming, and the need for the graduating chemical engineer to understand CRE principles became apparent.

Let's look at where chemical reaction engineering is now and how this was accomplished. Most universities have 15-week terms so consider the distribution of content material by the number of weeks devoted to a specific topic. Table 1 shows a typical distribution of the subject material at the University of Michigan. Discussions with colleagues reveal it is not too much different at other universities.

Table 1 - Typical Distribution of Material in an Undergraduate CRE Course.

1.	The Balance Equations	1 week
2.	Rate Laws, Activation Energy, and Stoichiometry	1 week
3.	Batch, Plug Flow, and CSTR Reactor Design	2 weeks
4.	Membrane and Semibatch Reactor Design	1 week
5.	Analysis of Rate Data	½ week
6.	Multiple Reactions in Batch, Plug Flow and CSTR Reactor Design	1 ½ weeks
7.	Reaction Kinetics and Mechanisms	½ week
8.	Enzyme Kinetics and Bio Reactor Designs	1 ½ weeks
9.	Heat Effects for Single Reactions, Multiple Steady States	2 ½ weeks
10.	Multiple Reactions with Heat Effects	½ week
11.	Gas-Solid Catalysis Laws, Mechanisms, and Decay	3 weeks
12.	Reactor for Catalyst Decay, STTR, Moving Bed	1 weeks
		16 weeks

There are various perturbations at different universities to the subjects and time allocations shown in Table 1. The most common topics that replace two or three weeks of material in Table 1 are shown in Table 2.

Table 2 - Variations of Topics Covered in CRE

A.	Diffusion and Reaction in Porous Catalyst	1 week
B.	Reactors for Catalyst Poisoning	1 week
C.	Mixing and Residence Time Distribution, Modeling, Non Ideal Reactors	2 weeks
D.	Chain and Polymer Reactions	2 weeks

For example, Topic 7 in Table 1 (Enzymes) would be replaced by Topic B in Table 2 (Poisoning).

A review of the material that was available in textbooks over 60 years ago surprisingly indicated that most of the subject matter in Tables 1 and 2 was discussed in volume three of Hougen, Watson and Ragaatz, *Chemical Process Principles*. The topics in Tables 1 and 2 that were missing or not covered in sufficient detail for an undergraduate course are shown in Table 3.

Table 3 - Topics Missing in the 1940s

1.	Derivation and Similarities of the Balance Equation
2.	Grater Understanding of Activation Energy
3.	Continuous Stirred Tank Reactors (CSTRs)
4.	Semibatch Reactors
5.	Membrane Reactors
6.	Moving Bed Reactors
7.	Straight Through Transport Reactors
8.	Algorithm for Handling Complex Reactions
9.	Rigorous Applications of the Energy Balance
10.	Multiple Steady States
11.	Enzyme Kinetics and Bio Reactors
12.	Multiple Reactions with Heat Effects
13.	Modeling Non-Ideal Reactors
14.	Gas Phase Reactions with Change in Total Number of Moles and Pressure Drop

2. Calculations in CRE

A. In the 1940s - Numerical Analysis and Slide Rules

The material was taught through involved example problems, many of which did not show how the end results were achieved. There were many times where gross approximations were made to facilitate slide rule calculations which used an Euler type integration scheme for differential equations,

$$X_{i+1} = X_i + f(X_1)\Delta Z$$

or perhaps a predictor corrector method. These calculations were all carried out with slide rules. It would take days to calculate temperature profiles for two or three different setting of the parameter or operating variables.

B. The 1950s

In the 1950s, many solutions to reaction engineering homework problems usually involved analytical solutions. However, a small number of universities assigned problems that could take advantage of mechanical type business calculators.

At the close of the 1950s, the tools for solving CRE problems are shown in Figure 1. The slide rule was the main calculational tool. The pencil and paper were used to record the calculations. When physical properties were needed, they were often obtained from nomographs. Rectangular, semi-log and log-log graph papers were used for plotting, and “French” curves used to “regress data” to smooth curves.

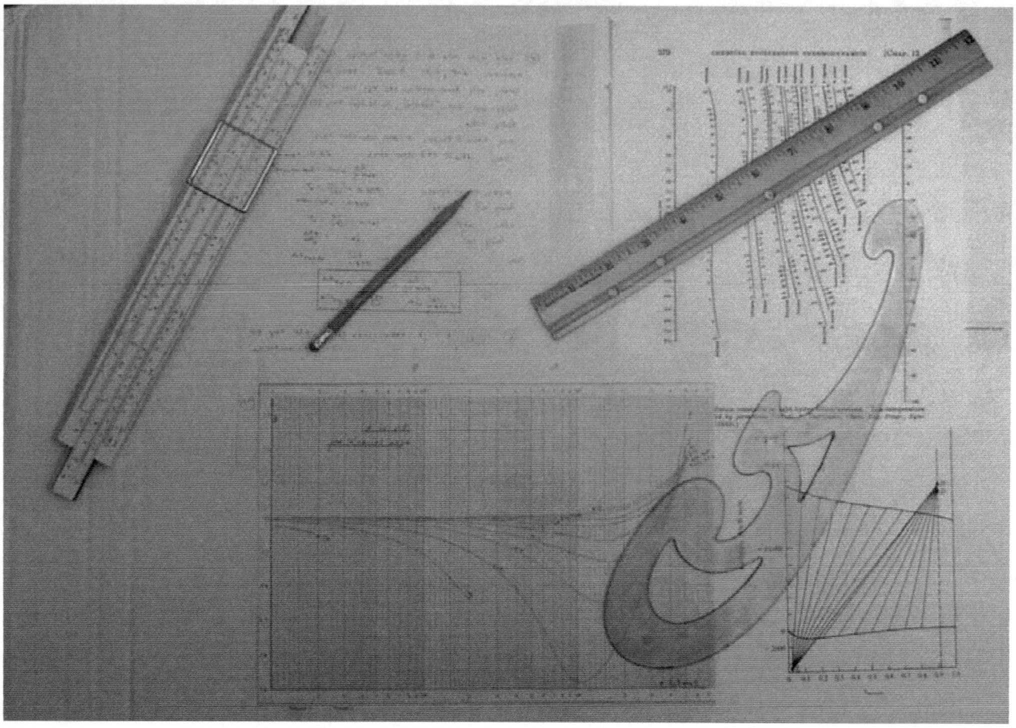

Figure 1 – CRE Student Tools in 1950s and Early 1960s

C. The 1960s

Analog and Hybrid Computers

The hybrid computers of this era used electronic circuits to perform calculations and were particularly useful for differential equations. The entire problem needed to be formulated into an electronic circuit using power supplies, function generators, initial condition power supplies, etc. The output was given to an oscilloscope or strip chart recorder. Figure 2 shows a Heath electronic analog computer kit which was used by some academic departments.

Figure 2 - Heath Electronic Analog Computer ~ 1960

Solutions of the problems involving series reactions, for example, required students to first wire up the problem and adjust the needed amplifiers and function generators to electronically simulate the problem and then "scale" the results back to the concentration variables of the reaction system. For example, in some exams or homework assignments, wiring sheets were used and the students were asked to show how to connect the wires to solve the a problem in a tubular flow reactor for reaction systems such as

$$A \rightarrow B \rightarrow C$$

Hybrid computers combined the analog computer with a small digital computer so that the analog problems could be under the control of a rudimentary digital computer. Some solid state electronics allows more accurate amplifiers so that more accurate results were obtained. All results were scaled to either ± 10 or ± 100 volts dc.

Digital Computers

Digital computers that were used in CRE at many universities started with the IBM 1401 and the much larger 360 series which followed. The programming language was usually FORTRAN that required programming for numerical solutions. Considerable effort was required to create a working program and make it adequate at solving CRE type problems. Users in the early 60s will remember when one run a day was the norm for developing and using FORTRAN programs.

Punched Cards

In the mid 60s, computer input was made with punched cards with some 72 columns of program information. Columns 72-80 were used to number the order of the cards so programs could be reassembled when the decks of cards were accidentally dropped. User of this era will remember that card punch machines that were used for hours to type up or correct the programs. Undergraduate CRE students of this period were using FORTRAN and other languages (such as MAD – Michigan Algorithmic Decoder) to write programs to solve reaction engineering problems with heat effects using a Runga-Kutta-Gill numerical scheme for differential equations. Hours and hours were spent as the students developed and keypunched his/her program and then waited in line for the output (if the program worked) to get the printout and his/her punched cards back. Most of the hours went into debugging the program thereby leaving little time to ***explore*** the problem. With the exception of the term projects, home problems that required only analytical solutions were assigned.

D. The 1970s

Punch cards continued to be used until the mid 1970s when FORTRAN programs could be written from computer work stations. Slide rules use disappeared completely as electron pocket calculators came into use in the mid 1970s for manual calculations. The first electronic calculator, the HP-35, is shown in Figure 3. This revolutionary calculator and others that followed allowed students to make desktop calculations such as log, ln, exponentials, trig functions, etc., and this allowed students and professionals to solve many problems much more rapidly.

Figure 3 – The HP-35 Electronic Calculator

Some CRE courses began using the IBM CSMP software (Continuous Systems Modeling Program) in the early 70s. This was a higher-level programming language that was suitable for integrating systems of ODEs on IBM System 360 mainframes. It also provided for easy control of initial conditions, repetitive integrations, inclusion of data from tables, and other convenient features. Output was either tables or very simple plots created from a standard alpha-numeric printer.

E. The 1980s

Computer programming, mainly with FORTRAN on large mainframe computers made by IBM, Control Data Corporation and others, continued to be used through the 1980s into the 1990s to solve complex CRE problems. Additionally, smaller computers such as the Digital Equipment Corporation (DEC) PDP series, were providing FORTRAN and other languages for making CRE calculations.

In 1981, IBM introduced the personal computer that was responsible for a dramatic increase in the availability of making CRE calculations on the desktop of faculty and students. Apple Corporation also pioneered the use of small personal computers. Again, FORTRAN compilers were initially used for programming of CRE problem solutions. Programs such as MATLAB[1] began to emerge that allowed for solution of numerical problems with increasingly less effort on the part of the user.

F. The 1990s

The next big advancement or paradigm shift in teaching undergraduate chemical reaction engineering came in the early 1990s with the introduction of inexpensive site licenses for POLYMATH[2] software by the CACHE Corporation[3] (Computer Aids for Chemical Engineering). This software was equation-orientated and provided interactive capabilities for the numerical solutions of CRE problems summarized in Table 4. Perhaps the most important was the ease in which systems of ODEs could be solved. Previous to POLYMATH, the students would spend 12-16 hours developing and testing FORTRAN codes to solve problems with heat effects and nonlinear reaction kinetics. After spending the time to defray the code, little time was left to explore the different parameters and operating conditions. With POLYMATH, one simply typed the ordinary differential equations (ODEs) as they were written on the paper, and they are solved immediately. Usually the programming took 15 to 30 minutes, after which the student was more than sufficient time to explore the problem and ask "What if..." questions. With POLYMATH it was also easy to address more complex problems such as multiple reactions with heat effects where some reactions are exothermic and some endothermic such as the production of styrene from ethyl benzene. Other situations that can be explored with POLYMATH in CRE are co-current and counter current heat exchange, membrane reactors, and bioreactors to mention a few.

Table 4 - Topics Easily Solved with POLYMATH

Multiple Reactions
Co and Counter Current Heat Exchanger
Membrane Reactors
Bioreactors
Multiple Reactions with Heat Effects
Nonlinear Regression of Rate
Solutions to Nonlinear Algebraic Balances on CSTRs

G. The 2000s until Today

Students and faculty in the current decade are routinely using interactive software, such as POLYMATH, Mathcad[4], and MATLAB, on their own personal desktop or notebook computers to solve problems in CRE that were not ever considered in the early years of this field of study. This allows faculty teaching CRE to focus on the development of CRE principles in many major new areas while the generation of the numerical solutions of the resulting mathematical models has become several orders of magnitude less than in the days of FORTRAN programming on mainframe computers. The important educational benefit is that the theory of CRE can continue to be developed while the numerical solution of the resulting mathematical problems becomes much less of a challenge.

It is interesting to look back briefly some 20 years into the "then and now" solutions of a CRE problem that was first handled by a FORTRAN program and compare the levels of effort required to achieve a numerical solution. Consider the modeling of a non-adiabatic tubular flow reactor for the catalytic oxidation of SO_2 first discussed by Fogler in the second edition published in 1986. A partial listing of the FORTRAN program is reproduced in Table 5. Note that a number of essential subroutines necessary for this program and for output generation are not shown. This program is very complex and has many opportunities for errors to be made before an accurate solution is finally obtained.

Table 5 – Partial FORTRAN Program for SO_2 Oxidation in a Non-adiabatic Tubular Flow Reactor

```
C       REACTOR SIMULATION FOR SO2 OXIDATION PROBLEM
C       PROGRAM WRITTEN BY A.K. DATYE
        DIMENSION T(1001),P(1001),X(1001),XEQ(1001),Z(1001)
        REAL KT,KP,KX
        COMMON /A/ KT(4),KP(4),KX(4)
      1 READ (5,1000,END=999) TO,PLOT,ITER
        TMAX= TO
        XMAX= 0.0
        DW=28.536/ITER
        I = 1
        T(I) = TO
        X(I) = 0.
        P(I) = 2.0
        Z(I) = 0.
        XEQ(I) = EQUIL(2.,TO)
C       BEGIN RUNGE-KUTTA-GILL ALGORITHM
C       ESTIMATE DERIVATIVES AT POINTS 1 THRU 4 AND
C       THEN COMPUTE AN AVERAGE DERIVATIVE FOR STEP I
        DO 100  I=1,ITER
C       DERIVATIVE AT START OF INTERVAL, POINT 1
        CALL DERIV(T(I),P(I),X(I),1)
C       DERIVATIVE AT STEP 2
        T2 = T(I) + KT(1)*DW*0.5
        P2 = P(I) + KP(1)*DW*0.5
        X2 = X(I) + KX(1)*DW*0.5
        CALL DERIV (T2,P2,X2,2)
C       DERIVATIVE AT STEP 3
        T3 = T(I) + 0.20710678*KT(1)*DW + 0.29289322*KT(2)*DW
        P3 = P(I) + 0.20710678*KP(1)*DW + 0.29289322*KP(2)*DW
        X3 = X(I) + 0.20710678*KX(1)*DW + 0.29289322*KX(2)*DW
        CALL DERIV (T3,P3,X3,3)
C       DERIVATIVE AT STEP 4
        T4 = T(I) - 0.70710678*KT(2)*DW + 1.70710678*KT(3)*DW
        P4 = P(I) - 0.70710678*KP(2)*DW + 1.70710678*KP(3)*DW
        X4 = X(I) - 0.70710678*KX(2)*DW + 1.70710678*KX(3)*DW
        CALL DERIV (T4,P4,X4,4)
C       COMPUTE CHANGE OVER STEP USING AVG VAL OF DERIV
        T(I+1) = T(I) + DW/6.*(KT(1) + KT(2)*.585786 + KT(3)*3.41424 + KT(4))
        P(I+1) = P(I) + DW/6.*(KP(1) + KP(2)*.585786 + KP(3)*3.41424 + KP(4))
        X(I+1) = X(I) + DW/6.*(KX(1) + KX(3)*.585786 + KX(3)*3.41424 + KX(4))
        Z(I+1) = Z(I) + DW/(1.4251432)
        XEQ(I+1) = EQUIL (P(I+1),T(I+1))
        IF (T(I+1).LE.TMAX) GO TO 100
        TMAX = T(I+1)
        XMAX = X(I+1)
  100   CONTINUE
        ITER=ITER+1
        WRITE (6,1001)Z(1),T(1),P(1),X(1),XEQ(1)
        WRITE (6,1002) Z(ITER),T(ITER),P(ITER),X(ITER),XEQ(ITER)
        WRITE (6,1003) TMAX,XMAX
  200   CONTINUE
        IF (PLOT.LE.0) GO TO 300
        DO 250 L=1,ITER
        X(L) = X(L)*4.0 + 2.0
        XEQ(L)=XEQ(L)*4.0 +2.0
        T(L) = (T(L)-1200)*0.01 +2.0
        Z(L) = Z(L)*0.2 +2.0
  250   CONTINUE
        CALL PAXIS(2.,2.,'DISTANCE ALONG REACTOR, FT',-26,4.,0.,0.,5.,0.5)
        CALL PAXIS(2.,2.,'FRACTIONAL CONVERSION',21,4.,90.,0.,0.25,0.5)
```

Table 5 (Cont.) – Partial FORTRAN Program for SO_2 Oxidation in a Non-adiabatic Tubular Flow Reactor

```
      CALL PAXIS(6.,2.,'TEMPERATURE DEG. RANKINE',-24,4.,90.,1200.,100.,0.5)
      CALL PLINE (Z,X,ITER,1,0,0,1)
      CALL PLINE (Z,XEQ,ITER,1,0,0,1)
      CALL PLINE (Z,T,ITER,1,0,0,1)
      CALL PLTEND
  300 GO TO 1
  999 STOP
 1000 FORMAT (2F10.0,I5)
 1001 FORMAT (//,' CONDITIONS AT BED INLET',/7X
     1,'Z',14X,'T',14X,'P',14X,'X',13X,'XEQ',/,5E15.7)
 1002 FORMAT (' CONDITIONS AT BED OUTLET',/,5E15.7)
 1003 FORMAT (//,5X,'MAXIMUM TEMPERATURE IN BED',1PE15.7,' DEG RANKINE',
     1/5X,'LOCATION OF MAXIMUM : ',F4.2,' FT. FROM ENTRANCE')
      END
      SUBROUTINE DERIV(T,P,X,I)
      REAL K1,K1P,KT,KP,KX
      COMMON /A/ KT(4),KP(4),KX(4)
C     SUBROUTINE FOR COMPUTING THE DERIVATIVES KX(I),KT(I),KP(I) RESP.
      F  = 912.84 -110.1*ALOG(T) -176008./T
      K1 = EXP(F)*3600.
      K1P = EXP(42311/(1.98*T) - 11.24)
      IF (X.GT.0.05) GO TO 10
      RATE = K1*(0.848 -0.1176/K1P**2)
      GO TO 20
   10 RATE = K1*((1-X)/X)**0.5*((0.2-0.11*X)/(2.-.11*X)*P
     1-(X/(1-X)/K1P)**2)
   20 KX(I) = RATE/0.188
      STCPI = 57.23 +0.014*T - 1.788E-6*T**2
      DELTCP = -1.563 + 0.00273*T -0.738E-6*T**2
      HR     = -42471 -1.563*(T-1260) +6.815E-4*(T**2 -1578600)
     1- 2.459E-7*(T**3 - 2.0003E9)
      KT(I) = (5.11*(1265-T) - RATE*HR)/(STCPI+DELTCP*X)/0.188
      VIS = 0.09*(T/1390.)**0.5
      FACTOR = 5500*VIS + 2291.1664
      KP(I)  = -1.1158334E-8*(1.-0.055*X)*FACTOR*T/P
      KP(I)=0.0
      RETURN
      END
      FUNCTION EQUIL(P,T)
      REAL K1P
      K1P = EXP(42311/(1.98*T) -11.24)
      XL = 0.
      XR = 1.0
      FL = 0.1*P*K1P**2
      FR = -1.055
C     INTERVAL HALVING METHOD
      DO 20 I=1,100
      XM = (XL+XR)*0.5
      FM = 0.11*P*(0.91-XM*.5)*(1.-XM)**2*K1P**2 -XM**2*(1.-.055*XM)
      IF (ABS(FM).LT.0.001) GO TO 40
      IF (FR*FM.LT.0.) GO TO 10
      XR = XM
      FR = FM
      GO TO 20
   10 XL = XM
      FL = FM
   20 CONTINUE
   40 EQUIL = XM
      RETURN
      END
```

The complexity of the partial FORTRAN program of Table 5 can be compared with the "Now" version of a complete POLYMATH program for this same problem as shown in Figure 4. Today's user can employ the interactive POLYMATH editor on a personal computer to easily enter the differential and algebraic equations. Syntax is automatically checked during entry, and the equations can be entered in any order as they are automatically ordered before solution. "If … then … else" logic can be utilized, for example, as is employed in the calculation of the reaction rate, **ra**, in Table 5. Robust integration algorithms with automatic step size adjustment and low error tolerances can be selected prior to problem solution. The user can select some combination of a Report, Graph, or Table to provide the results. For example, a graphical plot of the reactor temperature profile from POLYMATH is shown in Figure 5.

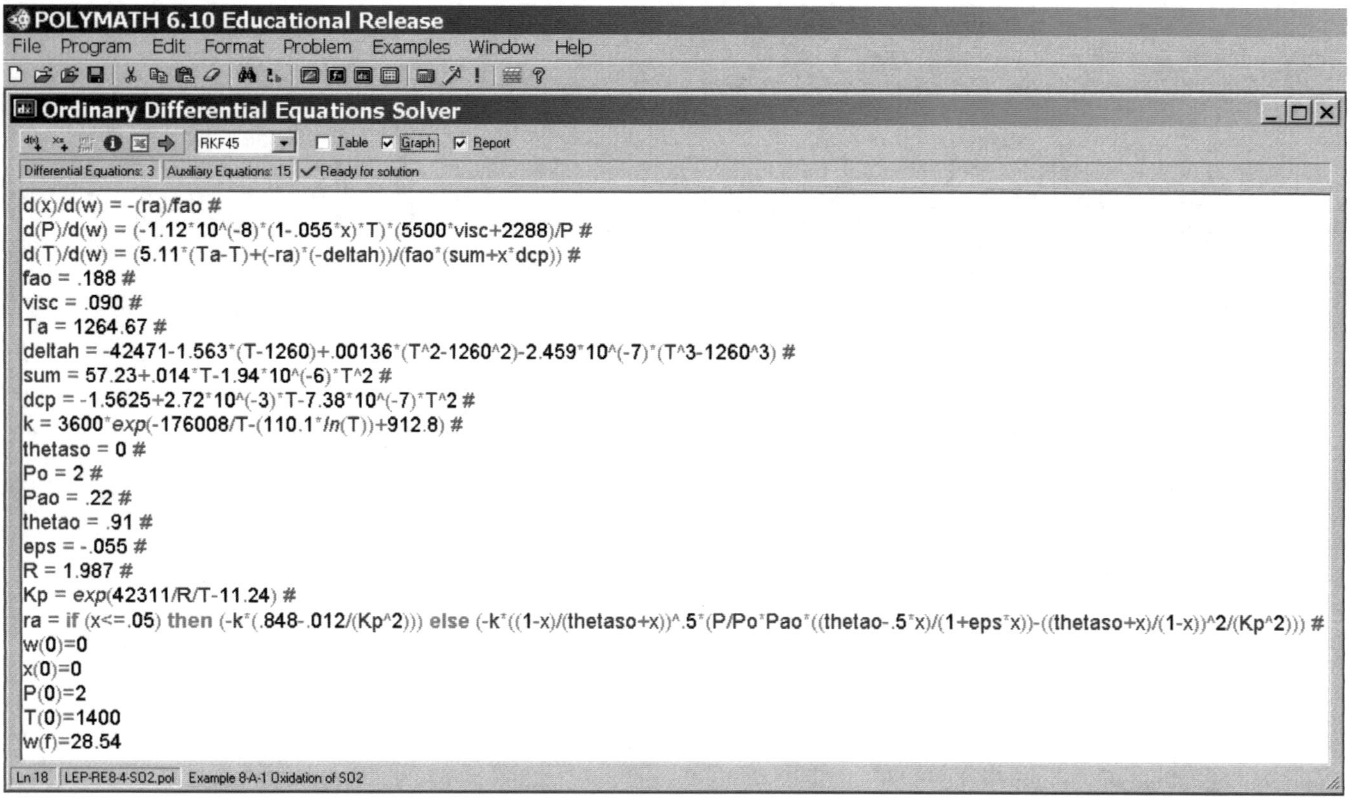

Figure 4 –POLYMATH Program for SO_2 Oxidation in a Non-adiabatic Tubular Flow Reactor

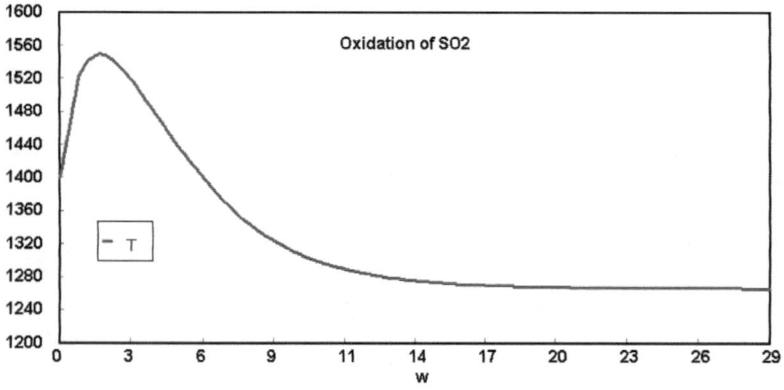

Note that POLYMATH can also automatically generate an Excel[5] spreadsheet solution for this problem and can also provide ordered MATLAB code for an m-file solution to this problem.

Figure 5 – Graph from POLYMATH Program for SO_2 Oxidation in a Non-adiabatic Tubular Flow Reactor

H. The Future

As CRE continues to develop, numerical solutions to partial differential equations that are used to model every more complex reactors and reacting systems will be needed. In this regard, software such and MATLAB and COMSOL[6] (formerly FEMLAB), for example, have developed software programs that can deal with PDEs. The current status is such that general use is coming but at present students can typically only change various reaction parameters in available solutions. Unfortunately, current students typically cannot easily change the model or system with the "canned" programs. However, if students learn how to program with software such as COMSOL, they can study any reaction in a tubular reactor. In the near future, it will probably be as easy to deal with PDEs as is now available for ODEs.

CRE appears to have a bright future in chemical engineering and related fields as it is well grounded in fundamentals that can be applied into the many developing areas of emerging and new technologies. It seems that the necessary computing and important software will continue to be developed that will allow students, faculty, and practicing engineers to solve ever more complex CRE related problems.

References

1. MATLAB is a registered trademark and software product of The Mathworks (http://www.mathworks.com/)
2. POLYMATH is a product of Polymath Software (http://www.polymath-software.com/)
3. CACHE Corporation (http://www.cache.org/)
4. Mathcad is a registered trademark and software product of the PTC Corporation (http://www.mathsoft.com/)
5. Excel is a registered trademark and software product of the Microsoft Corporation (http://www.microsoft.com/)
6. COMSOL is a registered trademark and software product of COMSOL, Inc. (http://www.comsol.com/)

Bibliography

Aris, R., and N.R. Amundson, *Mathematical Methods in Chemical Engineering Volume 2 First-Order Partial Differential Equations with Applications,* Prentice-Hall Inc., Englewood Cliffs, NJ 1973.

Aris, R., *Elementary Chemical Reactor Analysis*, Prentice-Hall Inc., Englewood Cliffs, NJ 1969.

Aris, R., *Introduction to the Analysis of Chemical Reactors*, Prentice-Hall Inc., Englewood Cliffs, NJ 1965.

Boudart, M., *Kinetics of Chemical Processes,* Butterworth-Heinemann Reprint Series 1991, originally published by Prentice-Hall in 1968.

Butt, J.B., *Reaction Kinetics and Reactor Design,* Prentice-Hall Inc., Englewood Cliffs, NJ 1980.

Butt, J.B., *Reaction Kinetics and Reactor Design, Second Edition, Revised and Expanded,* Marcel Dekker, Inc., New York 2000.

Carberry, J.J., *Chemical and Catalytic Reaction Engineering,* McGraw Hill, New York NY 1976.

Cooper, A. R. and G. V. Jeffreys, *Chemical Kinetics and Reactor Design*, Prentice-Hall Inc., Englewood Cliffs, NJ 1971.

Davis, M.E. and R. J. Davis, *Fundamentals of Chemical Reaction Engineering*, McGraw Hill, New York, NY 2003.

Denbigh, K.G., *Chemical Reactor Theory, An Introduction,* Cambridge University Press, 1965.

Fogler, H. F., *Elements of Chemical Reaction Engineering, Fourth Edition*, Prentice-Hall Inc., Upper Saddle River, NJ 2005.

Froment, G. F. and K. B. Bischoff, *Chemical Reactor Analysis and Design, Second Edition,* John Wiley & Sons, New York 1979, 1990.

Gray, P. and S. K. Scott, *Chemical Oscillations and Instabilities Non-linear Chemical Kinetics*, Clarendon Press, Oxford 1990.

Harriott, P., *Chemical Reactor Design,* Marcel Dekker, Inc., New York 2003.

Hayes, R.E., *Introduction to Chemical Reactor Analysis,* Gordon and Breach Science Publishers, 2001.

Hill Jr., Charles G., *Introduction to Chemical Engineering Kinetics & Reactor Design*, John Wiley & Sons, September 1977.

Holland, C.D. and R. G. Anthony, *Fundamentals of Chemical Reaction Engineering, Second Edition,* Prentice-Hall Inc., Englewood Cliffs, NJ 1989.

Hougen, O.A., and K.M. Watson, *Chemical Process Principles, Part Three: Kinetics and Catalysis*, John Wiley & Sons, New York 1964.

Levenspiel, O., *Chemical Reaction Engineering, Third Edition*, John Wiley & Sons, New York 1999.

Levenspiel, O., *Chemical Reactor Omnibook*, Oregon State University Book Stores, Inc. Corvallis, Oregon 1989.

Nauman. E.B., *Chemical Reactor Design,* John Wiley & Sons, New York 1987.

Nauman. E.B., *Chemical Reactor Design, Optimization, and Scaleup,* McGraw Hill, New York NY 2001.

Petersen, E. E., *Chemical Reaction Analysis*, Prentice-Hall Inc., Englewood Cliffs, NJ 1965.

Rase, H.F., *Chemical Reactor Design for Process Plants, Volume One: Principles and Techniques,* John Wiley & Sons, New York 1977.

Rase, H.F., *Chemical Reactor Design for Process Plants, Volume Two: Case Studies and Design Data,* John Wiley & Sons, New York 1977.

Rawlings, B. and J. G. Ekerdt, *Chemical Reactor Analysis and Design Fundamentals*, Nob Hill Publishing, Madison WI, 2002.

Riet, K. V. and J. Tramper, *Basic Bioreactor Design,* Marcel Dekker, Inc., New York 1991.

Smith, J. M., *Chemical Engineering Kinetics*, McGraw Hill, New York NY 1981.

Walas, S. M., *Reaction Kinetics for Chemical Engineers*, McGraw Hill, New York NY 1959.

258c From Ethanol and Butanol Fermentations to Cell- and Tissue-Culture ... and Back to Biofuels, 100 Years of Bioreactor Design, Operation and Challenges

Eleftherios Terry Papoutsakis, Dept. of Chemical Engineering & Delaware Biotechnology Institute, University of Delaware, 15 Innovation Way, Newark, DE 19711

Ethanol fermentations were established industrial process in late 19th and early 20th century prior to the formal genesis of chemical engineering, but the acetone-butanol-ethanol fermentation in the 1910s coincides with the evolution of chemical engineering as a distinct discipline. Chemical engineers however had their most profound impact in the mid 1940s in resolving the botttleneck for the development of large-scale, oxygenated, deep-tank fermentations for antibiotic production. This effort and the classical now work by Elmer Gaden and co-workers constitutes the foundation of biochemical engineering as a distinct sub-discipline. The late 1950s and 1960s brought to fore a wealth of new challenges and ideas with the concepts of bioreactors for immobilized enzymes and cells for biotransformations and whole-cell biocatalysis, in addition to the concept of computer-controlled fermentors. The 1970s saw an explosion of ideas on new bioreactor designs (including the various air-lift designs, many of which are now widely used), and numerous bioreactor-operation policies and their optimization including the now widely employed concept of fed-batch fermentations. The 1980s were marked by the realization that the design of bioreactors for growing animal cells (whether on microcarriers or in free suspension) was a new major challenge that required fresh analysis and new mixing concepts and scale up. This challenge was met with characteristic success leading to the growth of the now dominant animal-cell biotechnology for the production of protein therapeutics. This was followed by more specialized challenges at smaller scales to develop bioreactors for tissue engineering applications, and scaled-down technologies for the development of small or microbioreactors for parallel or high-throughput bioprocessing. Now, as we face the urgency to develop sustainable technologies for the production of chemicals and biofuels from renewable resources, there is a fresh set of challenges for efficient bioreactors and processes to meet perhaps the most diverse set of needs and applications.

258d Ultra-High Temperature Transport Tube Reactor Processing - from Electrical to Solar Heating

Alan W. Weimer, Department of Chemical and Biological Engineering, University of Colorado, 1111 Engineering Drive, Boulder, CO 80309-0424

This presentation will focus on the use of ultra-high temperature transport tube reactors for producing non-oxide ceramic powders and fuel intermediates. In both cases, reactors operate between about 1000 and 2000 C. The process to produce non-oxide carbides using resistive heating has been commercial for over 10 years. The process using concentrated sunlight to produce reduced metal oxides for water splitting thermochemical cycles and to convert biomass to sygnas will be commercialized sometime in the not too distant future. Discussion will focus on the key attributes of the reactors for both processes, including the challenges overcome to commercialize the ceramic powder process and the challenges that need to be overcome for the solar-thermal reactors to be commercial.

326a The Next Century of Process Principles...or Let's Just Settle for the Next Ten Years

Richard M. Felder, Chemical and Biomolecular Engineering, North Carolina State University, Raleigh, NC 27695-7905

In the past century, American chemical engineering students could pretty much map out their careers while still in college and be fairly accurate, at least for the next decade or two. They would graduate, go into the chemical or petroleum industry and be process or project or production or sales engineers, OR, go to graduate school and then take a research position in a chemical or oil company or join a faculty. If they were in industry, they would either continue doing what they did in their first position or move into management, and if they were on a faculty they would prepare other people to do the same things previous generations of chemical engineering students were taught to do and ended up doing.

That was a different world from the one we're inhabiting now. Chemical engineering graduates still go to work for chemical and oil companies but they're doing fewer traditional engineering tasks, many of which have now either moved offshore or been taken over by computers. Moreover, the graduates don't go into those two industries in nearly the numbers they used to--they're also going into biotechnology, biomedicine, nanotechnology, infotechnology, environmental engineering, medicine, and law, among other fields. Making things even more complex, the chances are that our graduates will change jobs fairly often in their careers, forcing them to continue acquiring new knowledge and skills—and even if they stay in the same place, the rapidly evolving capabilities of technology and shifting demands of the marketplace will force them to keep learning and growing. The traditional chemical engineering curriculum is poorly equipped to prepare our students for any of those things.

The stoichiometry course sets the tone for the chemical engineering curriculum and is the one course that most practicing chemical engineers say they continue to refer back to. This talk will address the following questions:

1. How can the stoichiometry course begin to equip students with the skills they will need to get and keep jobs in the coming decade? What changes in the current course content are required? What about the way the course is taught?

2. How should the stoichiometry course text evolve to keep pace with whatever changes are needed in the course?

3. What role should instructional technology play in all this? Can textbooks be replaced by software? Should they be?

326b Using Process Principles as An Introduction to Process Technology

Ronald W. Rousseau, School of Chemical & Biomolecular Engineering, Georgia Institute of Technology, 311 Ferst Drive, Atlanta, GA 30332-0100

The first course in a discipline is an eye-opener for new students: an initial interaction with concepts and core technologies, meeting faculty members with whom they may have continuing interactions, and an introduction to a unique language and perspective of the world they are entering. In chemical engineering (or chemical and bioX engineering) this comes together in what traditionally has been called Chemical Process Principles, Mass and Energy Balances, or simply Stoichiometry. However, development of a skill-based culture in which rigorous attention is given to quantitative analysis and the synthesis of processes based on a relatively small number of concepts is an essential, yet almost unacknowledged, feature of this course. These features were among the philosophical bases for the authors of the text *Elementary Principles of Chemical Processes* (R. M. Felder and R. W. Rousseau, Wiley, 3rd ed., New York, 2000.) and led to their emphasis of real chemical species in their examples and problem sets. It also was the rationale for including case studies in which students were presented descriptions of functioning chemical processes, and then led through those processes with series of problems that increased in depth and complexity. The importance of these issues in planning and teaching this essential course will be illustrated in the planned presentation.

326c Incorporating Process Synthesis into Introductory Chemical Engineering Courses

Regina M. Murphy, Chemical and Biological Engineering, University of Wisconsin, 1415 Engineering Drive, Madison, WI 53706

Several years ago the chemical engineering faculty at the University of Wisconsin-Madison decided that its introductory class, Chemical Process Calculations, was in sore need of an overhaul. In the traditional course, material and energy balance equations were taught and a few concepts in thermodynamics were introduced, and by the end of the semester, students, given a well-defined problem, could set up and solve steady-state process flow calculations. But they had little understanding of how these calculations were related to actually designing processes and making products. In redesigning the course, we set the goal of teaching students so that, by the end of the semester, given a poorly-defined problem with a minimum amount of information, they could synthesize a chemical process flowsheet that would approximate real industrial processes. This includes not only calculation of material and energy flows but also selection of separation technologies, determination of reasonable operating conditions, and integration of energy needs. We believed this would be possible at the introductory level through the use of limiting cases, approximations, and heuristics. Further, we believed this approach would motivate and engage students at a much higher level than the traditional. We also believed that this approach would greatly enhance students' understanding of the connection between the chemistry and the process, as well as their appreciation of the influence of economics, history, and geography on process design. Finally, we wanted students to learn that the principles of chemical process analysis, and the strategies of chemical process synthesis, can be applied to an enormous diversity of problems, from intracellular trafficking of a drug to accumulation of pollutants in the ecosystem. A new course led, by a long and winding road, to a new text, *Introduction to Chemical Processes: Principles, Analysis, Synthesis*, that was published by McGraw-Hill about two years ago. In this talk I will discuss some of the unique features of the text and present a few examples of the kinds of design problems that sophomore students have tackled in recent years.

326d A Historical Perspective of the Evolution of the Teaching of Process Principles

Jim Riggs, Texas Tech University, Dept of Chemcial Engineering, Lubbock, TX 79410 and David M. Himmelblau, Department of Chemical Engineering, The University of Texas at Austin, 1 University Station C0400, Austin, TX 78712-0231.

Process principles, primarily material and energy balances, have been the foundation of the teaching and the practice of chemical engineering since the inception of chemical engineering as a separate profession. Before World War II, process principles were empirically applied primarily using slide rules, mechanical calculators, graphs and tables. After World War II, a brief interlude of analog computers occurred, but their complexity to program (with wires) led to a more rational approach to solving material and energy balances that paralleled the development of digital computers. Today extremely large sets of material and energy balances are solved in the industrial application of process analysis, design, and optimization. Nevertheless, process principles, by in large, continue to be taught using 'hand solutions' of simple process equations, or their simplification for more complex processes, based on the argument that we need to avoid having students fill in dialog boxes, and instead provide students with a general understanding of material and energy balances for process systems. This presentation will focus on this evolution of the teaching of process principles from the early days based on precomputer methods (e.g., textbooks by Walker, Lewis, McAdams; Gilliland; and Hougen and Watson.) followed by the book of Henley and Rosen which was among the first to employ the more systematic approaches used today.

326e The First Course: An Introduction to Process Systems

Gintaras V. Reklaitis, Chemical Engineering, Purdue University, 480 Stadium Mall Drive, West Lafayette, IN 47907-2100

The traditional material and energy balance course serves many educational objectives in preparing students for the remainder of the curriculum and thus runs the risk of addressing a multitude of disparate topics. The narrower perspective taken in my textbook was that this course is the first exposure of the student to process flowhsheets and the rigorous accounting of material and energy flows and thus serves as the foundation for an integrated view of process systems. To build this foundation the course and book thus has three specific educational goals:

1. To provide a thorough exposition for balance equation concepts

2. To develop a systematic framework for the analysis of flowsheet information and specifications

3. To present systematic approaches for solution of flowsheet balance problems

In most ChE curricula, this integrated view of process systems is not revised until the process control and design courses in the senior year. All of the other courses, essentially fill in the detail with regard to the various unit operations "boxes" that the student sees in the first course. Thus, just as the design course is often described as the captstone course, the first course is can in fact be termed the cornerstone course for the process systems perspective.

440a Michael Cutlip, Panelist

Michael B. Cutlip, University of Connecticut, 191 Auditorium Rd., Rm. 204, Storrs, CT 06269-3222

Michael B. Cutlip is an emeritus professor and former chair of the Department of Chemical, Materials, and Biomolecular Engineering at the University of Connecticut. He is a coauthor of the widely-used POLYMATH software package. He has served as national programming chair and national chair of the Chemical Engineering Division of the American Society for Engineering Education and was the co-chair of the ASEE Summer School for Chemical Engineering in 2002.

440b Thomas Edgar, Panelist

Thomas F. Edgar, University of Texas at Austin, 1 University Station C0400, Austin, TX 78712

Thomas F. Edgar is the George T. and Gladys H. Abell Chair in Engineering at the University of Texas at Austin, and serves as the Executive Officer of CACHE (Computer Aids for Chemical Engineering) corportation.

440c Gintaras (Rex) Reklaitis, Panelist

G. V. Reklaitis, Chemical Engineering, Purdue University, Forney Hall of Chemical Engineering, Room 1019, 480 Stadium Mall Drive, West Lafayette, IN 47907-2100

Gintaras Rekalitis is the Edward W. Comings Distinguished Professor of Chemical Engineering at Purdue University. He has served as the Editor in Chief of the journal Computer & Chemcial Engineering since 1994 and is a past president of CACHE Corp. He was program chair for the 2002 AIChE Annual Meeting.

September 22, 2008 rmk

Integrating Modern Biology into the ChE DNA through a Campus-Wide Core Laboratory Education Program

Susan Carson[1,2], *John R. Chisnell*[1], *and Robert M. Kelly*[1,3] *

[1]Biotechnology Program
[2]Department of Plant Biology
[3]Department of Chemical and Biomolecular Engineering

North Carolina State University
Raleigh, NC 27695-7905

*Address correspondence to: **Robert M. Kelly**
Department of Chemical and Biomolecular Engineering
North Carolina State University
Raleigh, NC 27695-7905

Phone: (919) 515-6396
Fax: (919) 515-3465
Email: rmkelly@eos.ncsu.edu

INTRODUCTION

Efforts to infuse modern life sciences concepts into the traditional Chemical Engineering curriculum have accelerated in recent years, giving rise to a "new" sub-discipline, often referred to as *biomolecular engineering*. Chemical Engineering programs worldwide have embraced *biomolecular engineering* to the point that many departmental names now include this (or a related) moniker, reflecting a commitment to a permanent evolution of the discipline in this direction. One might argue that we have been here before (i.e., biochemical engineering), but the professional and technological driving forces to embrace biology within engineering have never been stronger. Our students are very interested in how biology fits into the Chemical Engineering discipline and how they can be prepared for professional opportunities in the biotechnology arena. As such, the task before us is to train students to function at the interface between Chemical Engineering and modern biology so that they can contribute in emerging areas ranging from biopharmaceuticals to biomaterials to bioenergy.

Truth be told, emphasis on the molecular life sciences (and other molecular sciences as well) has revitalized Chemical Engineering and will ultimately reinvent the discipline in many ways. However, with all this excitement comes a significant challenge – how to modify Chemical Engineering education so that its critical underpinnings are not slighted while, at the same time, going beyond a superficial treatment of the molecular life sciences. *In other words, Can a ChE "educational genotype" be developed that intrinsically and seamlessly integrates the life sciences?* This experiment is being currently being run at many institutions worldwide and requires rethinking of how Chemical Engineers are trained. To this end, many creative modifications of the undergraduate curriculum have been proposed and implemented to include biology. However, a major shortcoming typically arises: given the strong emphasis on sophisticated laboratory skills associated with the molecular biosciences and the lack of suitable lab courses and facilities on most campuses to deliver such training, how can an effective laboratory component be included? This is actually an issue not only for Chemical Engineering departments, but also for life sciences disciplines that need to provide laboratory-based training in modern biology for their students.

CORE LABORATORY TRAINING IN MOLECULAR BIOTECHNOLOGY

The expensive and sophisticated nature of laboratory techniques used in molecular biotechnology presents a challenge to those charged with developing pertinent instructional programs. In most cases, training along these lines on university and college campuses happens in a highly decentralized way: college-by-college, department-by-department or, for graduate students, research lab-by-research lab. These approaches seem to work at some level, but they suffer from several potential problems:

- The required effort and expense to carry out such training for particular programs can be prohibitive and, in the end, the results may be ineffective.
- In research labs, student-to-student "instruction" can propagate incorrect methodologies in addition to the fact that equipment, materials and supplies intended for research can be wasted as a consequence of the "learning curve".
- In isolated settings, interdisciplinary communication about new developments and advances in biotechnology techniques may not happen. Given the expanding array of disciplines, including biomolecular engineering, that can benefit from molecular biotechnology skills, decentralized training efforts will fail to capture this potential.

In any case, more effective strategies are needed to offset inadequate training in the range of lab skills associated with molecular biotechnology.

At North Carolina State University (NCSU), a core laboratory facility has been established and operates through its Biotechnology (BIT) Program (www.ncsu.edu/biotechnology) to teach molecular biotechnology skills to students campus-wide. This core facility was originally made possible by a unique funding arrangement between the colleges whose students benefited from the courses offered. Staffing was funded through a codicil between five colleges: Agriculture and Life Sciences, Engineering, Veterinary Medicine, Natural Resources, and Physical and Mathematical Sciences. The laboratory instrumentation and expendable supplies are covered through an allocation from a campus-wide educational technology fee, which allows the core facility to provide instruction in a modern laboratory setting with provision for continuous updating of instrumentation. *The overarching philosophy of the NCSU BIT Program is that molecular biotechnology encompasses a spectrum of theoretical knowledge, skills and lab techniques needed for advances in modern life science research and technology.* Molecular biotechnology is not an end in itself but rather a means to solve problems, unravel mechanisms and develop new technologies with societal benefit. The NCSU BIT Program is committed to enriching the base of scientific knowledge and laboratory skills necessary for genetic manipulation of living things at the molecular level. It also requires students to address the ethical issues surrounding biotechnology so that they can decide for themselves whether the merits of these new capabilities outweigh the associated risks.

CAMPUS-WIDE PARTICIPATION IN THE NCSU BIT PROGRAM

The NCSU BIT Program currently trains approximately 300 graduate and undergraduate students (as well as some postdocs, technicians, and faculty) annually, coming from over 35 Departments and Programs campus-wide. The BIT courses are very popular, and most sections fill each semester. From Fall 2001 through Spring 2008, there have been over 2,300 enrollments in BIT courses, representing students in 8 out of 10 of the colleges at NCSU. Non-matriculated students may take courses through Lifelong Education (LLE); in fact, part-time students currently employed with companies in Research Triangle Park are a growing component through a certificate program. Figure 1 shows the distribution of enrollments by college. Among graduate student enrollments, not surprisingly, 55% were from the College of Agriculture and Life Sciences (CALS), but approximately 17% came from the College of Engineering (COE). Among undergraduates, approximately 50% of enrollments were from CALS and 43% were from COE.

The overwhelming majority of students from COE were from Chemical and Biomolecular Engineering, at both the graduate (75%) and undergraduate (93%) levels (Figure 2).

ACADEMIC FRAMEWORK FOR MOLECULAR BIOTECHNOLOGY EDUCATION

While students can enroll in specific BIT courses to meet educational or research interests, training in molecular biotechnology is officially recognized through campus-wide academic minors and a certificate program. These are described below in brief:

Undergraduate biotechnology minor: open to all NCSU undergraduates, across all colleges, and all majors (see Table 1). The Department of Chemical and Biomolecular Engineering recently developed a biomolecular engineering concentration, which embeds the biotechnology minor into the B.S. degree. In addition to the normal ChE classes and BIT minor, students also take biochemistry and a course in biochemical/biomolecular engineering to earn the concentration.

Graduate biotechnology minor: open to all NCSU Master's and Doctoral degree candidates that have taken appropriate prerequisite courses and whose thesis research is in an area of molecular biotechnology (see Table 1). In addition to learning molecular biology techniques through lab courses, students have a unique opportunity to interact with peers in other disciplines. This has led to interdisciplinary research collaborations between groups on campus that may not have otherwise occurred. The interesting thing is that these have been student-driven. For example, food scientists working with veterinary medicine students to develop novel drug delivery methods and chemical biology and chemical engineering students developing novel ways to do microwave-driven biocatalysis. Such interactions would not have likely occurred without the BIT Program.

Graduate Certificate Program in Molecular Biotechnology: instituted in 2005 to provide post-baccalaureate students with the opportunity to obtain university-recognized credentials in molecular biotechnology. This Certificate Program is primarily geared toward non-traditional students who have already entered the workforce or are seeking to re-enter the workforce. For example, it provides an excellent opportunity for traditionally-trained, practicing chemical engineers to gain expertise in molecular biology lab skills. NCSU graduate students who are not eligible for the Graduate Minor because their thesis/dissertation work is not in an area of biotechnology are also eligible for the Certificate Program. Participating students must have completed all prerequisites (or their equivalent) to the Core Technologies course (or "Core Course", see below) before acceptance to the certificate program. The minimum requirements for the Core Technologies course are a general biology course that covers gene regulation and DNA replication, and two semesters of organic chemistry. A general microbiology course is recommended but not required. Prerequisites for the coursework electives vary, and those prerequisites may be taken after admission to the certificate program, if necessary. Requirements for completion of the Graduate Certificate Program are detailed in Table 1.

WHAT COURSES ARE OFFERED?

All students begin the program by taking a course entitled: *Core Technologies in Molecular and Cellular Biology*. This "Core Course" is a four-credit course offered in both Fall and Spring semesters that covers the basics of methods in recombinant DNA and protein expression, and serves as the prerequisite for all of our other courses. It is a semester long class that meets each week for a two-hour lecture and a five-hour lab. Students completing this course are awarded our annual BIT Program tee shirt. These usually are designed around one of our new modules and have slogans, such as "BIT happens" or "get bit" (see Figure 3). Following completion of the Core Course (or demonstrating prior equivalent training), students can take advanced laboratory courses, referred to as "modules," which have the same weekly two-hour lecture, five-hour lab periods, but are only 2 credits and last only one-half semester. Students are able to mix and match modules according to their interests. Most of these have only the Core Course as a pre-requisite – this structure creates sufficient flexibility for students campus-wide to pursue the graduate and undergraduate minors. Most of the courses are offered at both graduate and undergraduate levels. In general, students who take courses for undergraduate credit are expected to master techniques and concepts covered in the course, and to be able to analyze data and trouble-shoot experiments gone awry. Students taking the courses for graduate credit have the additional expectation of being able to develop their own experiments and protocols in the area covered in the course. Advanced undergraduates are able to take graduate-level courses by special permission. Brief description of course content is provided Table 2.

WHO TEACHES THE LAB COURSES?

Since we are a Program and not a Department, one of the most common questions asked is: "Who teaches these courses?" The answer is that instructors are recruited from across campus and are at different career stages, but they have certain things in common: 1) they are volunteers and not required to teach our classes; 2) they have expertise in the area of instruction, both conceptual and at the bench; 3) they have a strong desire and ability to teach sophisticated modern biology lab techniques effectively. The majority of our instructors are tenured or tenure-track faculty from a variety of departments and colleges. Their home departments receive "credit hours taught" for the classes they offer, thus the instructors' Department Heads are typically supportive of the arrangement. Some of our instructors are Research Faculty or Research Associates. These are individuals with doctoral degrees whose appointment at the university is primarily research, but who have interest in developing their teaching skills to prepare them for academic appointments. They teach courses on a semester-by-semester basis. We currently have one Teaching Assistant Professor whose appointment is to teach for the Biotechnology Program and to serve as academic coordinator. Finally, we employ one to two teaching postdoctoral fellows (see below).

This instructional model has allowed us to maintain instructors with a high level of energy, expertise, and enthusiasm. One of the benefits to teaching in our program is that all of the students are there because they want to be, thus creating a very stimulating educational environment. Graduate teaching assistants are typically recruited from the labs of the instructors – these are the people who use the techniques and technology on a daily basis and bring a very positive and essential element to the courses. Course evaluations are routinely very positive – *no doubt the product of interested instructors and teaching assistants teaching interested students.*

CHALLENGES FACING CAMPUS-WIDE LABORATORY INSTRUCTION

Campus-wide programs focusing on life sciences face some unique challenges due to their interdisciplinary nature. Discussed below are some that we have faced.

Pre-requisites: In a given class in the NCSU BIT Program, we have students from majors ranging from chemical engineering to plant biology to textiles to microbiology to veterinary medicine. Because we are a campus-wide core laboratory education facility and our mission is to train students from diverse disciplines, it is critical that the prerequisites for our courses not be restrictive. The only prerequisites to our Core Course are a freshman level biology course that covers DNA replication, transcription and translation at a very basic level, and chemistry through the second semester of organic chemistry (2 semesters of general chemistry and 2 of organic). For this reason, background knowledge varies considerably from student to student. Despite this potential problem, the courses have gone surprisingly well. Learning to appreciate how other disciplines approach technical issues can be an education in itself.

Clicker technology: Students with engineering backgrounds tend to have an easier time with quantitative aspects, while students from life sciences majors, such as microbiology and genetics, typically have a better background in biological concepts. We have worked to overcome these differences in a variety of ways. One method we have used in the classroom is the implementation of *clicker technology*. Interspersing clicker questions throughout each lecture in our Core Course allows students to work through questions on their own, without the risk of one student calling out the answer before each student has a chance to think it through. It also helps students not understanding a concept to weigh in on the pace of the instruction without fear of being embarrassed by asking a "simplistic" question.

Aseptic technique: In the lab, in a given semester, perhaps half of the students have already had microbiology with microbiology lab (or cellular biology), and thus only about half have skills in aseptic technique. Time is limited in our core course so that proper coverage of aseptic technique at the level it is covered in general microbiology lab is not possible. We overcome this limitation by partnering students in the lab such that different majors are represented; students can learn from each other. We also limit lab groups to two students so that significant participation is expected from everyone.

Graduate/Undergraduate Mix: Almost all of our courses are dual-level courses, with undergraduate and graduate students in the same classroom and lab. Because many of our courses are resource-intensive "boutique" courses, and because undergraduates and graduate students do need the same set of skills, it makes sense from a budgetary standpoint to combine the levels to reap the greatest benefit from the equipment available. Generally, the students in the graduate and undergraduate level courses have the same lectures and the same laboratory experiences. However, as mentioned previously, the graduate students are expected to perform at a higher level of understanding and proficiency than the undergraduates, as assessed by special assignments given to the graduate students. For example, all students would be expected to apply concepts to solve problems, analyze and trouble-shoot experimental data, and design some of their own experiments, but graduate students might be expected to write a mini-research proposal, designing experiments from start to finish, or to evaluate a journal article for its scientific merit. Although we had initial concerns about grouping graduate and undergraduate students, we have not experienced any significant problems in delivering our courses – not surprising since the lab skills being taught are typically new to everyone.

NEW AND CONTINUING INITIATIVES

The core laboratory education model has also served as a nucleation point for many campus-wide efforts that go beyond our instructional program and a test bed to try new things which would not be possible in a typical departmental setting. Some examples of such activities are discussed below.

Undergraduate Research: There is no doubt that direct involvement in research greatly enhances the undergraduate educational experience (5, 8). The process of researching a topic in the primary literature, designing experiments, implementing those experiments, and analyzing the results is critical for developing the analytical skills necessary to function in the biotechnology sector. The benefits of an undergraduate research experience is reflected in increased undergraduate graduation rates (7), increased rates of pursuit of graduate education (4, 3), and increased interest in science careers (2). The National Academy of Sciences strongly recommends that all undergraduate students be given the opportunity to participate in original research (6). To earn the undergraduate biotechnology minor, students must participate for at least one semester of research on a molecular biotechnology-based project or work for a period of time with a biotechnology company. With respect to the latter, the Research Triangle Park provides ample opportunity for such an experience. We also have two new initiatives directed an enhancing the undergraduate research experience.

First, partnering with the Department of Plant Biology at NCSU, we have received an NSF Research Experience for Undergraduates (REU) grant in the area of synthetic biology in plant systems. The program commences in Summer 2009 and will allow students from universities outside of NCSU to pursue summer research experiences on our campus. All students in the summer program will first complete "biotech boot camp", an accelerated, intensive version of our Core Course, so that they have the proper skills to work effectively in the labs they choose to join on campus for the summer. We think that this preparation will make the REU experience better for both students and mentors since some preparation in molecular biology skills will precede their laboratory experience.

Second, we are in the process of developing an inquiry-guided research course for freshman. In the proposed course, students will isolate naturally-occurring bacteriophage (viruses that infect bacteria, but not humans) from the environment. They will culture, isolate, and titer the bacteriophage (phage), visualize the phage by transmission electron microscopy, and then purify its genome. Students will determine whether the phage has a double-stranded or single-stranded DNA or RNA genome, and then determine the size (number of nucleotides) of the genome. The culmination of this project will be to prepare and present a poster for our annual campus Undergraduate Research Symposium. Our goal is to have students experience the scientific process first-hand early in their college education. This course would be open to students in all majors, with no college-level prerequisite. For non-science majors, the understanding of the scientific method may be more valuable than the facts they could learn in a traditional lecture course. For science majors, we anticipate that involvement in research projects will enhance their performance in science curricula, and will give them confidence to pursue more independent research earlier in their academic careers.

Teaching Postdocs: We also have implemented a Teaching Postdoctoral Program. This is a "win-win" program that gives recent PhD graduates, preferably with a conventional postdoc experience already completed, that also have a strong interest in an academic career, the opportunity to gain significant mentored teaching experience. Since the fellows come "fresh off the bench" with experience in the most cutting-edge technologies, they add an innovative dimension to our program. This fellowship allows for the postdoc to teach a section of the Core Course under the mentorship of an experienced instructor, and then develop and implement his or her own specialized half-semester course. We have "graduated" two teaching postdocs to date, and are working with our third. The two postdocs who completed the program have successfully acquired faculty positions at primarily undergraduate institutions. The new courses that have been added to our curriculum through this program are Mutagenesis, RNA Interference and Model Organisms, and a new course in Protein-Protein Interactions (including the yeast two-hybrid system).

Graduate and Undergraduate Training: The Biotechnology Program also oversees two externally-funded training programs at NCSU and contributes to several others. At NCSU, we have an NIH Biotechnology Training Grant and a Department of Education Graduate Assistance in Areas of National Need (GAANN) Fellowship Program in Molecular Biotechnology. Graduate students supported by these awards complete the graduate Biotechnology minor as well as benefit from other courses that we offer. A new NIH Training Program in Translational Medicine in the Veterinary College will also use elements of our courses for their students. The Microbial Biotechnology Master's program in the Department of Microbiology, which has a strong business focus, incorporates our molecular biology lab courses into their programs. Undergraduate students receiving biomanufacturing training at NCSU through the new Golden LEAF Biotechnology Training and Education Center (BTEC) and the Bioprocessing Science (BBS) major in the Department of Food, Bioprocessing and Nutrition Sciences integrate offerings from the BIT Program into their curricula.

SUMMARY

Thinking "outside the box" can sometimes lead to very effective educational outcomes. Though not without unique challenges, a campus-wide core educational program at NCSU has proven to be an efficient and effective way to deliver cutting-edge molecular biotechnology laboratory courses to students from a wide range of disciplines, including Chemical and Biomolecular Engineering. Housing expensive equipment in a single facility where it can be taken full advantage of, rather than duplicating equipment for teaching purposes in multiple departments, has been a clear advantage in the current budget climate. Furthermore, unforeseen advantages, such as interdisciplinary student collaborations, have arisen. We encourage other institutions to investigate whether this model will work for them and are happy to discuss our experiences with whoever might be interested.

With respect to integrating molecular biosciences into Chemical Engineering (the new "ChE genotype", if you will), laboratory training at sufficiently sophisticated levels that reflect state-of-the-art techniques and skills is crucial. While it may be infeasible for Chemical Engineering departments to provide this training on their own, partnerships campus-wide cannot only provide such opportunities but also create a new paradigm for training our students in the molecular sciences. As we re-think curricular design with biology in mind, it must be done with an eye towards the world of science and technology that our students will be entering: one that is interdisciplinary and dynamic, and one that will care not only about what they know but also what they can do.

REFERENCES

1. **Carson S. and D. Robertson.** 2005. Manipulation and Expression of Recombinant DNA: A Laboratory Manual, Second Edition. Academic Press.

2. **Fitzsimmons, S. J. and A. Associates.** 1990. A preliminary evaluation of the Research Experiences for Undergraduates (REU) Program of the National Science Foundation. Program Evaluation Staff, Office of Budget and Control, National Science Foundation.

3. **Hathaway, R. S., B. R. A. Nagda, and S. R. Gregerman.** 2002. The relationship of undergraduate research participation to graduate and professional education pursuit: An Empirical Study. Journal of College Student Development **43:**614-31.

4. **Kremer, J. F. and R. G. Bringle.** 1990. The effects of an intensive research experience on the careers of talented undergraduates. Journal of Research and Development in Education **24:**1-5.

5. **Lopatto, D.** 2004. Survey of undergraduate research experiences (SURE): First Findings. Life Sciences Education **3:**270-277.

6. **Morgan, E. and the Committee on Undergraduate Biology Education.** 2003. BIO2010: Transforming undergraduate education for future research biologists. National Academies Press.

7. **Nagda, B. A., S. R. Gregerman, J. Jonides, W. von Hippel, and J. S. Lerner.** 1998. Undergraduate student-faculty research partnerships affect student retention. The Review of Higher Education **22:**55-72.

8. **Seymour, E., A. B. Hunter, S. L. Laursen, and T. DeAntoni.** 2004. Establishing the benefits of research experiences for undergraduates in the sciences: First findings from a three-year study. Science Education **88:**493-534.

Table 1. Requirements for NC State Biotechnology Minors and Certificate				
Requirement	**Description**	**Undergrad Minor**	**Graduate Minor**	**Graduate Certificate**
General Biology (or equivalent)	Introductory biology course that includes coverage of gene regulation (i.e., lac operon).	X	X	X
Organic Chemistry	Two semesters of organic chemistry	X		
Manipulation of Recombinant DNA (Core Coure)	Intensive molecular cloning laboratory course (undergraduate level)	X		
Core Technologies in Molecular and Cellular Biology (core)	Intensive molecular cloning laboratory course (graduate level)		X	X
Advanced biotechnology laboratory modules	Students choose from courses described below	X (2)	X (1)	X (2)
Research experience	150 hours of biotechnology research on campus or other molecular biology-related research (academic, government or industrial)	X		
Thesis or dissertation	Thesis or dissertation research must utilize molecular biotechnology skills.		X	
Faculty representative	At least one member of the student's thesis/dissertation committee must be a member of the Biotechnology faculty.		X	
Lecture electives	Upper-level lecture course covering molecular biology topics. Examples: Prokaryotic Molecular Genetics, Plant Molecular Biology, Biochemistry of Gene Expression, Genetic Data Analysis, Functional Genomics, and Biochemical Engineering.			X
Ethical Issues in Biotechnology	Ethics and real-world issues in biotechnology and professional ethics.	X		X

TABLE 2. Course Offerings in NCSU Biotechnology Program	
Core Technologies of Molecular and Cellular Biology*	Sub-cloning a gene into an expression vector; screening for positive transformants by a variety of methods including DNA preparation, ligation and transformation, PCR, restriction mapping, colony hybridizations with DNA and monoclonal antibody probes, SDS-PAGE, and Western blotting; and purification of recombinant protein by affinity chromatography.
RNA interference and Model Organisms	History and application of RNAi technology; design of experiments to silence gene expression in various organisms; assessing extent of silencing; use of online tools for design of RNA silencing constructs to knockdown mammalian protein expression; advantages and disadvantages of model organisms; proficiency using *Nicotiana benthamiana* tobacco plants, *Caenorhabditis elegans*, and mammalian cell culture.
Phenotypic Analysis of Transgenic Plants	Phenotypic parameters that can be measured to characterize a mutant or transgenic plant line; methods and technologies that can be used for these characterizations.
Genetic Engineering of Eukaryotic Microbes	Importance of filamentous fungi and yeast in biotechnology and as research tools; manipulation and growth of these eukaryotic organisms 'in vitro'; genetic transformations of fungi; creation and analysis of mutant strains; expression of heterologous proteins in yeast.
Fermentation of Recombinant Microorganisms	Small-scale fermentations of recombinant *Escherichia coli* and *Saccharomyces cerevisae*; factors affecting gene expression and protein production.
Protein Purification	Chromatography techniques for protein purification, including ion exchange, hydroxyapatite, hydrophobic interaction, gel flitration, affinity; purification tables constructed based on SDS-PAGE analysis, enzyme assays, and protein concentration.
Protein-Protein Interactions	Basic concepts and techniques involved in the study of protein-protein interactions, including the yeast-2-hybrid system, pull-down assays, and immunoprecipitation.
Proteomics	Introduction to the theory and practice of proteomics, analysis of microbial proteomes, statistical data analysis, MS fundamentals.
Mutagenesis	Site-directed mutagenesis by a variety of methods in multiple organisms
Plant Tissue Culture and Transformation	Basic techniques in plant tissue culture and transformation in model plant species and agriculturally important plants.
Animal Cell Culture Techniques	Culture of embryonic stem cells; establishment and maintenance of large-scale eukaryotic cell culture for protein production.
Advanced Animal Cell Culture	Culture of embryonic stem cells; establishment and maintenance of large-scale eukaryotic cell culture for protein production.
Genome Mapping	Basic techniques in genetic and physical mapping; principles of DNA marker development, marker detection, genetic and physical mapping; DNA sequencing
DNA Microarrays	Array design and printing; principles of data analysis and data mining using data acquired from actual experiments; importance of controls and statistical significance; global controls of gene expression.
RNA Purification and Analysis	Isolation of RNA and quantification by spectrophotometry, and analysis by gel electrophoresis; northern blotting and non-radioactive labeling and detection by chemiluminescence.
Real Time PCR	Real-time PCR theory, techniques, machinery, troubleshooting, tools, and advanced protocols, such as multiplexing and SNP analysis
Ethical issues in Biotechnology **	Discuss and debate controversial topics in biotechnology

Computer Analysis of DNA Sequences **	Databases (particularly NCBI/GenBank) for finding homologs of genes and proteins of interest; tools in commonly used DNA analysis software packages.
Research Ethics **	Seminar/discussion of research ethics topics including authorship, animal/human subjects, bioethics, intellectual property, and research fraud
Capstone Biotechnology **	Molecular biotechnology-related research seminars by academic/industrial speakers; stock market competition; interdisciplinary group design project
Professional Development **	Seminar/discussion of topics related to career development, including public speaking, grant writing, CVs, post-doc opportunities, attending scientific meetings, interviewing skills.
**Manipulation and Expression of Recombinant DNA: A Laboratory Manual, 2E* was developed for this course (1) ** Lecture format only	

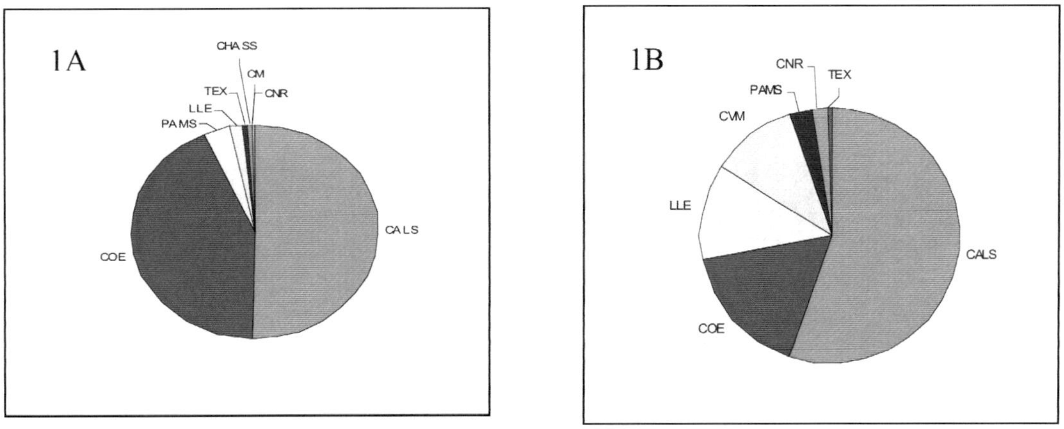

Figure 1. Student enrollments by college. Abbreviations: CALS = College of Agriculture and Life Sciences; COE = College of Engineering; CVM = College of Veterinary Medicine (graduate program, only); PAMS = College of Physical and Mathematical Sciences; CNR = College of Natural Resources; TEX = College of Textiles; CHASS = College of Humanities and Social Sciences; CM = College of Management; LLE = Lifelong Education (non-matriculated students)

1A. Percent of graduate student enrollments by college. CALS 55%; COE 17%; LLE 12%; CVM 10%; PAMS 3%; CNR 2%; TEX 1%.

1B. Percent of undergraduate student enrollments by college. CALS 51%; COE 43%; PAMS 4%; LLE 1%; TEX 1%, CHASS <1%; CM <1%; CNR <1%.

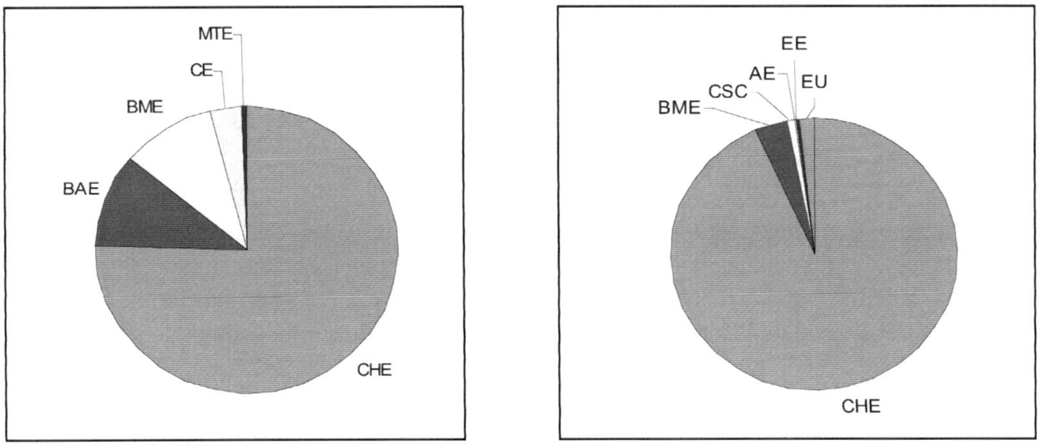

Figure 2. College of Engineering Enrollments. Abbreviations: CHE = Chemical Engineering; BME = Biomedical Engineering; BAE = Biological and Agricultural Engineering; CE = Civil Engineering; MTE = Materials Science And Engineering; EU = Engineering Undesignated; AE = Aerospace Engineering; EE = Electrical Engineering.

2A. Percent of different engineering majors among engineering graduate students. CHE 76%; BAE 11%; BME 10%; CE 3%; MTE <1%.

2B. Percent of different engineering majors among engineering undergraduate students. CHE 93%; BME 4%; EU 2%; CSC 1%; AE <1%; EE <1%.

Figure 3. Biotechnology (BIT) Program T-Shirts. Students successfully completing the Core Course earn the current year's version of the NCSU BIT Program tee shirt. From left to right: Dr. Scott Witherow, teaching postdoc; Rosemary Le, current chemical and biomolecular engineering undergraduate; Dr. Sue Carson, Biotechnology Program Teaching Assistant Professor and Academic Coordinator; Diana Bisbee, chemical and biomolecular engineering graduate (now employed at Biogen Idec); Melissa Cox, Biotechnology Program laboratory manager.

513b Biology In Che Education: Future-Research Perspectives

Michael L. Shuler, Chemical and Biomolecular Engineering and Biomedical Engineering, Cornell University, 120 Olin Hall, Ithaca, NY 14853-5201

Biology has changed tremendously in the last 25 to 30 years from a largely descriptive science to one with a strong foundation based on underlying molecular mechanisms. This change facilitates the ability of chemical engineers to make contributions to both a better understanding of how living organisms function and to the application of biological insights to problems of societal importance, such as human health, sustainable development, and cleaner environment. Biological mechanisms can be analyzed using the classic undergraduate ChE tools of stoichiometry, kinetics, thermodydnamics, mass transfer, and applied mathematics. Chemical engineers are well-positioned to address the key issue in biology: relating genomic and molecular insights to the physiology of whole organisms. However, students need additional background beyond the traditional ChE courses; primarily in biochemistry, cell biology, molecular biology, and physiology. When chemical engineers are involved in applications, the emphasis may change, but the same engineering and biological fundamentals pertain.

It is important to recognize that biology is more than new vocabulary; it is a different way of thinking. What constitutes "proof" in biology is radically different from what chemical engineers might consider proof. The "hypothesis-driven" research model of biology and the associated experimental approaches differ from typical chemical engineering education. A successful biochemical engineer needs to be able to appreciate the strengths in both the chemical engineering approach to problem solving and the biological approach to discovery. A biochemical engineer needs to synthesize these two intellectual frameworks into a coherent whole. A PhD biochemical engineer will develop this integrated, coherent perspective. At the undergraduate level, it is difficult to achieve this fusion of intellectual approaches, especially through coursework. Undergraduate, hands-on, research is the best approach for undergraduates to achieve a coherent perspective of biochemical engineering.

The opportunities for biochemical engineers in the next 50 years are great, but success will depend on educational experiences that achieve a true coherence in engineering and biological approaches to new knowledge.

513c Biology in Che Education: Future – Industry Perspectives

William J. Kelly, Villanova University, 800 Lancaster Ave., Villanova, PA 19085

The session will focus on how to prepare graduating engineers for the increasing role that biology plays in industry; including but not limited to the biopharmaceutical manufacturing. Regulatory and environmental constraints may require not only chemical engineers to have some knowledge of biology. Consequently, many engineering colleges are currently grappling with the questions like: do some or all engineers need some college-level biology training and which if any engineering departments should require this? Has the changing of engineering department names to include for example "biomolecular" been indicative of a shift towards more required or elective biology courses for engineering students, or in general more/better training in biology? The necessity of specific biology courses and topics in preparing engineering students for the industry of today and tomorrow will be discussed.

513d What'S What for Faculty Seeking Ready-Made Bio Curriculum

Claire Komives, Chemical and Materials Engineering, San Jose State University, 1 Washington Sq., San Jose, CA 95192-0082

The demand for college graduates who are skilled in applying chemical engineering principles to biological applications is growing. At the same time, students are eager to embark upon life science-based careers. Traditional chemical engineering education, with a focus on petroleum and commodity chemical processing, appears to be due for a facelift. However, few resources are available to facilitate faculty to make the shift towards incorporation of life science applications. Likewise, enticing and enabling faculty who have not had any formal training in biotechnology to participate in the bio movement can be thorny. This brief talk will summarize the materials currently accessible to faculty who wish to add more bioX into their courses, including a review of their utility. A handout will be provided to the attendees summarizing the list described in the presentation.

513e Biology in Che Education: Classroom and Laboratory Experiments in the University of California, Davis Curriculum

David E. Block, Department of Chemical Engineering and Materials Science and Department of Viticulture and Enology, University of California Davis, One Shields Ave, Davis, CA 95616, Karen A. McDonald, Department of Chemical Engineering and Materials Science, University of California, Davis, One Shields Ave., Davis, CA 95616, and Marjorie L. Longo, Chemical Engineering and Materials Science, University of California, Davis, One Shields Ave, Davis, CA 95616.

The Department of Chemical Engineering and Materials Science at UC Davis has had a distinct Biochemical Engineering major/program for over twelve years. As part of the curriculum for this major, several courses were created or expanded to parallel courses in the traditional chemical engineering curriculum. In addition to a fermentation course and a bioseparations course, students in the major take a course on biopharmaceutical manufacturing facility design and a biochemical engineering laboratory course, along with a senior design project focused on the production of a biological molecule. After an overview of this unique program at Davis, we will first focus this presentation on the course activities of the biopharmaceutical manufacturing course. These activities include a quarter-long group design project for a specific piece of equipment, in-class and in-pilot plant "hands-on" demonstrations of specific pieces of equipment, and highly-directed tours of commercial development and manufacturing facilities. Next, we will discuss the biochemical engineering laboratory course which includes a series of bench scale experiments covering both fermentation-based and downstream-processing-based activities. Finally, we will discuss how successful laboratory experiments from the biochemical engineering curriculum have been adapted for the traditional chemical engineering curriculum at Davis.

531a NSF and Engineering Education

John Daniels, Engineering Education and Centers, National Science Foundation, Directorate for Engineering, 4201 Wilson Boulevard, Arlington, VA 22230

The US National Science Foundation has been integrally involved in the larger national discussion on engineering education, the history of which is connected to the Engineering Education Coalitions of the 1990s and reports from the National Academies (e.g., Gathering Storm, Educating the Engineer of 2020). Likewise, the NSF provides leadership as to what sort of programs and investments ought to be pursued in the context of the American Competitiveness Initiative, the America Competes Act and general federal policy. This presentation will provide an update on NSF's perspective on a variety of initiatives (including department level curriculum reform) as well as general funding opportunities.

Web Page: www.nsf.gov/div/index.jsp?div=EEC

531b Curriculum Reform In the Systems Area

Thomas F. Edgar, University of Texas at Austin, 1 University Station C0400, Austin, TX 78712

A chemical engineering education should provide skills that graduates can use to become successful problem solvers in the process industries. Graduates should be able to

(1) Integrate knowledge and information to aid in solution of chemical engineering problems.

(2) Make estimates and assumptions, face open-ended problems, deal with noisy data and uncertainty, and envision possible solutions.

(3) Hone their problem-solving skills; use computational tools; perform economic analysis; and plan, execute, and interpret experiments.

The systems component of the chemical engineering curriculum ensures that chemical engineering graduates should be able to create and understand mathematical descriptions of physical phenomena; scale variables and perform order-of-magnitude analysis; structure and solve complex problems; manage large amounts of messy data, including missing data and information; and resolve complex and sometimes contradictory issues of process design. Graduates should be able to handle sensitivity of solutions to assumptions, uncertainty in data, what if questions, and process optimization. The systems approach is a fundamental concept that is explicitly addressed only in a few chemical engineering courses. The concept of analyzing a collection of components and processes as an overall system, rather than as individual components, is critical for newer areas of chemical and biological engineering, as well as for traditional areas. The knowledge base of systems consists of methods for dynamic and steady-state simulation at multiple length and time scales, statistical analysis of data, sensitivity analysis, optimization, parameter estimation and system identification, online monitoring and diagnosis design and analysis of feedback systems and design of products and processes.

T6011 Department Level Curriculum Reform

Two innovations: five-year program in Engineering and Management and the MPS program

Donald R. Woods
McMaster University
Hamilton Ontario Canada

Two unique curricular approaches have been taken at McMaster University.
In the Five-year *Engineering and Management* program students graduate with a fully-accredited degree in engineering plus the core courses of Bachelor of Commerce degree (micro and macro economics, finance, accounting, marketing, policy, organizational behavior, operations research); this is more than the first year of an MBA, plus unique Engineering and Management courses developed with industry on communication, people skills and project work for an external client that requires both commerce and engineering expertise. Although the program requires five years, the trick was to provide the required 42 credit hours of MBA material plus 6 units of commerce electives but only require one additional year of 38 credit. This was achieved by astutely providing credit within both programs and by spreading the courses carefully through the five year program. The first graduates were in 1975. The response by alumni and recruiters has been extremely positive. We have restricted enrolment in the program.

The MPS program was created in response to the need to improve our students skill in problem solving. We did an extensive five year research program to determine what we might do. We discovered that additional skills besides problem solving were needed (including time and stress management, interpersonal skills, conflict management, learning skills, goal setting, creativity, analysis, critical thinking, self awareness of uniqueness, lifelong learning, self assessment and so on). We found that workshops were the most effective vehicle to develop such skills. In the McMaster Problem Solving (MPS) program about 120 hours of workshops were spread over four required courses in the four year program (one course in sophomore year, two in junior year and one in senior year). The program was implemented in 1982. Details are given of the MPS program and its effectiveness at http://www.chemeng.mcmaster.ca/innov1.htm and MPS.

A common concern of Departments is how to best prepare students for their careers. Sometimes this dilemma is expressed as "to what extent should we include the business and management issues in our program?" Don't many engineering graduates take an MBA anyway? Another dilemma is "to what extent should we develop our students *process skills* in such areas as problem solving, team work, interpersonal skills, self assessment and lifelong learning. After all graduates of our program are good at problem solving because in a typical four-year engineering program they solve more than 3000 homework problems (1)". Semiconductors, bio, pharmacy, product design, energy, environment, safety, green engineering - the list goes on about different specializations that we might include in our program. How do we handle this, and more importantly, how do we anticipate such trends because it takes about five years of development work and getting students through our program before we can produce graduates with these "specialty skills"?

We too at McMaster University have tried to address these issues. Our response has been to develop a unique five year dual degree program in Engineering and Commerce called the Engineering and Management program. Secondly, we implemented the McMaster Problem Solving program to develop process skills and skill in lifelong learning that hopefully obviates the need to have many specializations to meet the ever-changing demands.

In this paper we describe first the Engineering and Management program; then the MPS program.

1.What about dual-degree programs in five years? McMaster's Engineering and Management Program

Here we consider the rationale, the challenges and opportunities, the Commerce core, efforts to integrate the engineering and business elements across the curriculum, integrating professionalism with academia via the Industrial Advisory Council, comparing the Engineering and Management graduate with a four-year engineering graduate plus one year of an MBA, and the response to the program. In Section 1.8 we describe how this five-year model has been extended to other areas at McMaster University.

1.1 Rationale

In the late 1960 our faculty realized that there was a need for engineers with the core ideas from Commerce that make up the first year of an MBA. The vision was for a five year program that blended of all the required courses for an accredited degree in Engineering with the core courses in a Bachelor of Commerce degree. Rather than calling this a combined degree in Engineering and Commerce, a new title Engineering and Management (B.Eng Mgt) was selected. The graduates of the program are fully accredited Engineers in each of the Engineering programs: Chemical, Computer, Electrical, Engineering Physics, Materials, Mechanical, Mechatronics and Software.

1.2 Challenges and opportunities

Such a program could not have been successful on our campus without the whole hearted cooperation between the Faculties of Commerce/Business and the Faculty of Engineering. We wanted the engineering students to take the commerce courses along with the commerce students and we wanted to program to be available to students in all of the engineering disciplines.

First, to prevent the commerce courses from being overloaded there were enrolment restrictions; the program was limited to 75 students.

Next, to fit the core Commerce into the one-year time, there had to be negotiations. The Bachelor of Commerce program has a required core of 42 credits. To achieve the integration with the required Engineering programs of each of the engineering disciplines without loss of parts of either the Commerce core or the full Engineering components, the Mathematics and Science courses are credited for both programs. Thus, 21 credits of Mathematics needed in the Commerce program are provided by the Mathematics taken in the Engineering program. In addition 12 credits out of the 18 credits of Humanities electives required by Engineering accreditation are allocated to Organizational Behavior, Economics and Commerce courses from the core of the Commerce program. The resulting integration maintains all the required components of both the Engineering and Commerce degrees.
Admission to the program is after the freshman year and by application.

Another challenge, once the program was approved, was time-tabling. The core Commerce courses had to be available to engineering students from all different branches of engineering. The Engineering and Management courses had to be available to students of all engineering disciplines.

The Director of the program was selected for 3 year term and alternated from being selected from the Faculty of Business and the Faculty of Engineering. The Director was expected to attend Faculty meetings in both faculties and to take a strong interest in both faculties.

1.3 The Commerce core

The core Commerce courses include 12 credits of economics (introduction, intermediate price theory, intermediate income and employment theory); 6 credits of organizational behavior (introduction and either industrial relations or personnel); 6 credits of marketing (introduction and market research); 6 credits of accounting (financial, cost and management accounting); 6 credits of finance (financial instruments and institutions; management finance); 6 credits of management science (production and operation, business policy) plus two electives.

All the courses are taken with students registered in the Commerce program. There are not separate courses for the engineers.

1.4 Integrating Engineering and Management courses

To integrate the two disciplines and to develop esprit de corps among the students in the five year program for all engineering disciplines we have a variety of Engineering and Management courses. These courses have evolved over the years. Because of my familiarity of the courses when I was director, these are the ones described here.

Two non-credit Engineering and Management seminars in years two and three. Activities included critiquing Tom Peters videotapes, fundamentals of communication, personal uniqueness, self performance analysis, law and career planning. The seminars could ask for no homework They had about 90% attendance and were extremely popular and useful in developing esprit de corps and providing skill development often missed in other formal programs.

Two one-credit, week-long Engineering and Management skill development workshops after years three and four. The first series of workshop focused on problem solving, creativity and decision making; the second workshop considered interpersonal skills, conflict resolution, networking, interacting with the media, dealing with difficult behaviors and creating mission and vision statements. The workshops were from 9am to 5pm for five days and were presented by Human Resource personnel from companies and from our faculty. They were typical of in-house workshops. They were conducted the first week after exams. The arrangement was that students would be hired by a company and paid to attend the workshops. In return the company could send to the workshop, free of charge, one employee for every student. This was very successful win-win situation.

Two Engineering and Management Communications courses. Students wrote reports on their summer work experience and how it relates to the career plans and the theory they have studied. The written reports are marked and discussed individually with the students by both faculty and professionals from the Industrial Advisory Council. Speeches are videotaped and reviewed privately.

Final year Engineering and Management project. Teams of 10 to 15 students from different engineering disciplines work on interdisciplinary projects from an industrial client that require project definition, market surveys, financial considerations, engineering calculations and project planning and scheduling. The results are presented orally to the client. Clients included Eldorado Resources (how best to satisfy our energy needs); Corundol Environmental (how to use current facilities to best respond to future trends); Joseph Brant Memorial Hospital (critique and recommendations for energy utilization); Wallace Barnes (increase production efficiency and market capacity for one of its product lines).

1.5 Integration of professionalism with academia via the Industrial Advisory Council

One of the major keys to success is the hard working industrial advisory council. The IAC are mid to senior level engineers who are managers from a cross-section of industries. Besides offering guidance and advice about the program, the Council members are vital to three and sometimes four educational components. First, each is a mentor to 15 students. They take an interest in the student's progress, provide insight about their organization and provide field trips. Each council member marks the student's written reports and personally goes over the reports with each student. Council members suggest projects for the projects course. Occasionally, they interview first year students applying for admission into the Management program. They are looking for leadership qualities.

The council usually meets several times a year to advise the Director and the Operating committee about the program.

1.6 Comparison between the Engineering and Management graduate and four-year engineering graduate plus one year of an MBA

The five advantages are:
1. The E&M graduate has taken 16 more credits of Commerce than one would take in the first year of an MBA.
2. The commerce and engineering courses are integrated throughout the program instead of being considered separately in sequence.
3. The interaction with the Industrial Advisory Council and the students and the mentor program helps the students understand how organizations function.
4. The E&M program provides a double and perhaps a triple emphasis on "communication skills": perhaps from within the engineering discipline, from within the commerce program and from within the E&M program.
5. The Commerce and Engineering courses are carefully planned so that there is no overlap. However, some engineers taking their first year of the MBA may be required to take mathematics courses where they have had "almost all of the math before."

1.7 Response to the Program

Based on the response from industry and the fact that most graduates receive a position by the time they graduate, this is a very successful program. Some companies come on campus just to recruit E&M students.

1.8 McMaster's other five-year programs

At McMaster we offer a five-year *Engineering and Society* program and a five-year Chemical Engineering and Bioengineering degree.

Like the Management program, the Society program is open to students in all engineering disciplines. The purpose of the Society program is to explore the human side of engineering and the complex interactions between technology and society. The program includes Engineering and Society focus electives plus integrating Engineering and Society courses that address the history, culture and social control of technology. http://www.eng.mcmaster.ca/engandsoc/
Within the Society program is the Engineering and International Studies program to help future engineers to better understand the complexities associated with global project management. The emphasis is on an understanding of and a sensitivity to the different cultural, political, religious

and historical backgrounds of potential collaborators, as well as additional technical training in areas that are particularly important in international settings, such as international project management and supply chain management.

Unlike the Management program, the five-year Chemical Engineering and Bioengineering degree is only open to students in Chemical Engineering. This program is a unique 5-yr program that combines the core, accredited Chemical engineering undergraduate curriculum with courses from the life sciences (biology, human anatomy and physiology, biochemistry of macromolecules, cellular and molecular biology) as well as integrating courses on bioreactors and bioseparations. http://chemeng.mcmaster.ca/programs.html

2. The McMaster Problem Solving Program

Consider first the rationale. Then we consider visualizing the curriculum, identifying the core fundamentals, the set of skills, the target behaviors of the process skills and the associated learning objectives, and how to develop the process skills; designing the workshops and the course sequences, and implementing and evaluating effectiveness.

2.1 Rationale

In the 1970s we realized two things: a) feedback from industry suggested that our graduates did not have skill in problem solving and b) the job market wanted some graduates in environmental engineering and some in semiconductors, in the former specialty we offered perhaps one or two courses but in the latter we offered no courses. This could be expressed as two generic questions: 1) how do we develop our student's abilities in the *process skills* (that some call soft, generic, higher order thinking or procedural skills) needed by professionals in their career? These might include such skills as problem solving, communication, team work, performance review, listening, time and stress management, lifelong learning and conflict resolution.
2) How can we offer a program that will equip graduates to work in the wide variety of industries where the basics of chemical engineering are needed: such as pulp and paper, polymers, semiconductors, environmental, pharmacy, foods, fertilizers, petrochemicals, refining, and tar sands.
The glib answer to both these questions is to develop our students skills in problem solving and in lifelong learning. If the graduates have confidence and skill in lifelong learning then we don't need to offer specialized courses beyond the fundamentals. No matter what the professional expectation is in their career they can learn and apply the new information if they know the fundamentals and can apply lifelong learning. Fortunately at that time McMaster Medical School had started its revolutionary program of small group, self-directed, self-assessed, interdependent problem-based learning. This was a unique and effective way to develop lifelong learning skills. Fortunately, we attended their workshops and learned much about their approach. However, this Medical School approach was challenging for us to implement because for the Medical School a) one admission criteria included performance in a PBL session that required skill in problem solving and team work (whereas students in most programs were admitted based on primarily on marks), b) students admitted to the medical school were usually graduates of

other undergraduate programs (and were therefore three to four years more mature than our engineering students), c) the whole program was PBL so that one faculty tutor could be assigned to a group of five students (whereas in engineering one faculty member would have a class of about 30 to 100 students) and d) most of the material was developed in the context of health sciences. Hence, attractive as this pedagogical approach might be, major modifications were needed to make it effective in our engineering classrooms.

2.2 Visualizing the curriculum

We realized that we couldn't convert our whole program to PBL; we hoped that we could have some senior level courses that could use PBL and we hoped that that exposure was sufficient to develop the required confidence and skill in lifelong learning. In previous courses we would focus on developing the process skills to make PBL work with autonomous, small groups of 5 such that 10 to 15 autonomous student groups would work as effectively as if they had a tutor present in the group. Self-assessment and self awareness thus became major skills to develop in addition to the problem solving and team skills. The steps to take included identifying the core fundamentals, learning how to develop the process skills, identifying the target behaviors of the process skills, designing the workshops and course sequences and implementing and evaluating the effectiveness.

.

2.3 Identifying the core fundamentals

Whereas senior students tend to see Chemical Engineering as 2000 equations that they have to memorize, we wanted them to be soundly based in the seven core fundamentals: conservation of mass, species, momentum, energy and charge and fundamentals of phase equilibrium and chemical reaction equilibrium. We visualized, for example, the conservation of energy via a concept map illustrating the extension to thermodynamics, fluid mechanics, heat transfer and reactor design. Copies of this map can be obtained from the author at <woodsdr@mcmaster.ca>. Scrutiny of the courses we taught suggested that about 2/3 of each course content was prerequisite for subsequent courses and about 1/3 was interesting enrichment that was not required subsequently. The result was that we could free-up about 12 credit hours to focus on the process skills to build up to the use of PBL, and that the core fundamentals could be addressed through courses before the beginning of the senior year. This would leave the senior year open for "graduate level" electives through which students could elect to experience the research specialists of the faculty.

2.4 Identifying the set of skills

It might sound obvious that the skills a professional requires includes problem solving, creativity, decision making, team work, and interpersonal skills. But how might we break those down into a subset of skills. Where do listening skills fit in? Is it important to include translation skills to convert a set of words into a diagram or do our students possess that skill already? To identify the skill set (and perhaps how to and how not to teach them) we received funding for a four year project during which a faculty member became a student again, attended all the required courses with the students. He met weekly with a cohort of the students to discover how they were 'solving the ordinary homework problems", learning how well they could extract the fundamentals from their courses and create a diagram of the structure of the knowledge and how it interacted between the Chemistry, Physics, Mathematics and other courses they were taking. From this we could determine details about the skill set needed by the students. I became a student again in 1974 and followed the same cohort through the four year program. I "graduated" in 1978. Dr C.M. Crowe became a student again in 1975 and followed another cohort for one year to validate the results I had observed and to provide a check on the skill set.

2.5 Identifying that target behaviors of the process skills and the associated learning objectives

We need to base our program on the fundamentals of cognition and not on the intuitive approach published in "how to do it" books. Fortunately there is a rich set of *novice* versus *expert* research from which we can extract the target behaviors (2). From such targets we created observable, learning objectives with measurable criteria (3) for each of the skills and subskills.

2.6 Identifying how to develop the process skills

From the above-mentioned research project we met with the cohort of students as we followed them through the four year program. During those meetings we would try different options to help them improve their skill; we invited different experts to observe and suggest how we might develop the skills. We discovered that, for example subject knowledge (of lack thereof) often interfered extensively with the problem solving process. It was crucial to set up situations that would help us discern whether the fault lay with a lack of knowledge or lack of problem solving skill. From this we learned to adjust our program to provide workshop experience on subject-independent content. In addition, we sat in on about half a dozen industrial in-house workshops to develop these skills. Bandura (4) provided insight as to how change skills and attitude via his research on how to change phobias. He found that observing someone successfully handle the phobia was ineffective. He recommended that the whole process be divided into small sections that can be handled easily, that the person be given a chance to try the experience where there was an extremely high probability of success. The person would receive feedback and celebrate success. So the skill and confidence would be built layer by layer.
We planned early versions of some of the workshops and then tried them with volunteers and with professionals from the Instructional Development Centre as observers (5).

2.7 Designing the workshops and the course sequences

As Bandura has illustrated, workshops, that give participants a chance to try a skill and get prompt feedback, seem to be the most effective approach. Here are the steps we used.

Step 1. Define the skill base on research, not intuition. Publish target behaviors of successful persons performing the skill. Some examples are published:
http://www.chemeng.mcmaster.ca/innov1.htm and click on MPS and look for "target skills"

Step 2. Create learning objectives for the skill. Some examples are published:

http://www.chemeng.mcmaster.ca/innov1.htm and click on MPS and look for "objectives"

Step 3. Design the worksheets (feedback forms and forms of evidence that participants will use to show performance accomplishments)

Step 4. Work out the "timing sheets" for how you will manage the workshops.
An overall template for the workshops is as follows:

Define
Rationalize
Pretest
Objectives
Route ahead
Activity in subject-free context (eg puzzles, reasoning) with feedback
Reflections
Activity with feedback
Description of target behaviors
Activity with feedback
Reflections
Activity with feedback
Continue until the participants feel they can achieve the objectives
Objectives
Post Test
Discovery

The first activities are in subject-free contexts so that the participants can focus on **"building"** the skill. The next activities are to "**bridge"** the application to the subject domain. Here we ask for the same thinking behaviors but apply them in the context of Chemical Engineering. Finally, the participants are ask to "**extend"** the application of this behavior to their homework in any discipline or course and to their everyday life.
Reflection and self-assessment about the process and the skill acquisition are critical components to the process.

The formal required course sequence was:

48 hours of workshops in the sophomore year where the focus is on individual skill in problem solving, self assessment, time and stress management, personal uniqueness, creativity and analysis.

18 hours of synthesis and application of the previous skills in the first semester junior year.

24 hours of workshops in the second semester junior year where the focus is on knowledge structure, interpersonal skills and team work.

30 to 50 hours of workshops in the first semester senior year on solving ill-defined problems such as trouble shooting, systems thinking, chairperson skills, and learning engineering economics using PBL (6).

By the senior year, we were confident that our students possessed the required problem solving and team skills that they would be admitted into the Medical School PBL program via the simulated tutorial described above under Rationale. Hence, that was when we felt the students, working in tutorless groups, were ready for PBL.

2.8 Implementing and evaluating effectiveness.

The revised program was implemented in 1982. The PBL sessions worked so well in the senior year that the students requested that PBL also be used as part of a junior level course. Currently PBL is used in both a third and a senior course. Along the way we had to address issues different from the medical school because we were using autonomous or tutorless groups (7).

We published a major evaluation of the program (8) and specific articles related to problem solving (9); team work (10) and lifelong learning and PBL (11).

The MPS program continues, with some modifications; and attracts numerous visitors and sabbatical leave colleagues.

3. Summary

Two successful innovations were described: the five-year Engineering and Management program and the McMaster Problem Solving program. The rationale for these is given. Steps taken to design and implement these are listed. The overall evaluation of these approaches is given.

4. References

1. Woods, D.R. (1985) "Problem-based Learning and Problem Solving," in "Problem-based Learning in Education for the Professions," D. Boud, ed., Higher Education Research and Development Society, Sydney

2. Woods, D.R. (1988) "Novice vs. Expert Research suggests ideas for implementation", PS Corner, J of College Science Teaching, vol.18, Sept., 77- 79; 66-67; vol. 18, Nov., 138-14; vol. 18, Dec/Jan., 193-195.

3. Visit Chemical Engineering Department's website http://www.chemeng.mcmaster.ca/innov1.htm and click on MPS

4. Bandura, Albert, (1982) "Self-efficacy mechanism in human agency" American Psychologist, 1982 Feb Vol 37(2) 122-147

5. Woods, D.R., C.M. Crowe, T.W. Hoffman and J.D. Wright, (1980) "What is the Problem in Teaching Problem Solving?" in "The Teaching of Elementary Problem Solving in Engineering and Related Fields," J.L Lubkin, ed., ASEE publications, Washington, DC., p 35-55.

6. Woods, D.R. (1994) "Problem-based Learning: how to gain the most from PBL," D.R. Woods, Waterdown, distributed by McMaster University Bookstore, Hamilton, ON, Canada

7. D.R. Woods, W.C. Duncan-Hewitt, F.A. Hall, C.H. Eyles and A.N. Hrymak (1996) "Tutored versus Tutorless Groups in Problem-based Learning," American Journal Pharmaceutical Education, 60, 231-238

8. Woods, D.R., A. N. Hrymak, R.R. Marshall, P. E. Wood, C. M. Crowe, T.W. Hoffman, J.D. Wright, P. A. Taylor, K. A. Woodhouse, C.G. Kyle Bouchard (1997), "Developing Problem solving skills: the McMaster Problem Solving program," J. Engineering Education, April, **86**, 2, 75-91.

9. D.R. Woods, T Kourti, P.E. Wood, H. Sheardown, C.M. Crowe and J.M. Dickson (2001), "Assessing Problem Solving Skills, 1: The context for assessment," Chemical Engineering Education, Fall, 35, 4, 300 - 307 and D.R. Woods, T Kourti, P.E. Wood, H. Sheardown, C.M. Crowe and J.M. Dickson (2002), "Assessing Problem Solving Skills, 2 assessing the process of problem solving," Chemical Engineering Education, Winter. Vol 36, no. 1 p 60 - 67

10. D.R. Woods S. Taylor and S. Jaffer (2005) "Assessing Team Work. part I: the group's performance," in Preparing for PBL. 3rd ed.; and D.R. Woods, S. Taylor and S. Jaffer (2005) "Assessing team work, part II: individual contributions to the team," in Chapter 8, Preparing for PBL. 3rd ed. This is downloadable free from the web http://www.chemeng.mcmaster.ca/innov1.htm and click on "Preparing for PBL"

11. D.R. Woods and Liz Bayley (2006), "Assessing student performance in Problem-based learning" Chapter 8 in *Preparing for PBL* 3rd edition This is downloadable free from the web http://www.chemeng.mcmaster.ca/innov1.htm and click on "Preparing for PBL"

The Pillars of ChE: An Integrated Curriculum at Pittsburgh

Joseph J. McCarthy[1], *Robert S. Parker*[1], *and Mary Besterfield-Sacre*[2]
[1]*Department of Chemical and Petroleum Engineering*
[2]*Department of Industrial Engineering*
University of Pittsburgh
Pittsburgh, PA 15261

Abstract

In this paper, we discuss the development, implementation, and assessment of a novel integrated curriculum in chemical engineering. This curriculum, based on block scheduling, consists of 6 main or "pillar" courses, augmented by a number of concurrent laboratory classes. The goal of this new curriculum is to foster integrated thinking beginning as early as the sophomore year, while at the same time enabling instructors to have the flexibility to utilize modern pedagogical tools. By using a world-class suite of assessment techniques including concept maps and concept inventories, we hope to produce a truly validated success story that can serve as a model that is applicable for all engineering disciplines.

1 Introduction

The National Science Foundation (NSF) has funded a number of engineering coalitions [1] to study "best practices" in engineering education. Overwhelmingly, these coalitions have favored active-learning activities and integration of complementary subject materials - often combining mathematics, physics, and chemistry in freshman engineering programs [2], for example. While these integrated curricula have been used successfully at a number of institutions [3], this approach has only rarely been used outside of the first year, and, to our knowledge, had never been adapted to a full engineering curriculum. In our project, we develop an integrated curriculum that spans the upperclass years, from sophomore to senior. Our fully integrated chemical engineering curriculum is unique for its use of block scheduling [4] – a technique with a strong literature base and proven track record in K-12 education – for the first time in a traditional higher education engineering curriculum. Block scheduling, in its simplest form, is transforming multi-semester courses into a single-semester course via extended, concentrated contact time.

Adapting these two proven educational methodologies has resulted in the 6 *Pillars of Chemical Engineering* [5]. These courses have considerably longer contact hours than a traditional university course so that: (1) students may gain systems insight through integration [6] of their core knowledge across traditional course and discipline boundaries; (2) the instructors have the time to include truly multi-scale (from molecular to continuum to macroscopic) descriptions of chemical engineering content; and (3) the instructors have the flexibility to accommodate diverse learning styles and incorporate active learning more effectively [7].

This new curriculum has been in use, and under active assessment, in the Department of Chemical and Petroleum Engineering at the University of Pittsburgh since 2002. In this paper, we first outline the concept and implementation of the Pillars curriculum. We follow this discussion by describing our assessment strategies and quote several interesting results. We complete the paper with an outlook to the future for this curriculum and others like it.

2 Pillars

Current engineering instruction is often compartmentalized within a traditional 3-4 credit per course schedule, so that knowledge is disconnected and well-defined relationships are established across a curriculum only during the senior year, if at all [8]. By moving to a block-scheduled curriculum, we have integrated complementary subject-matter along with experiments and open-ended problems, so that students see connections across the discipline *during each course.* From a logistical standpoint, the pillar courses are 5-6 credits each with an associated 1 credit lab for each, so that contact time is roughly 2 hours per day, 5 days per week. The goal of each pillar is to be a microcosm of chemical engineering, so that each course serves to enhance a student's understanding of the discipline as whole: in many cases attacking both theory and practical application in the same course. In this way, students build their concept of the discipline gradually rather than as an "epiphany moment" in their senior year, once they have all the

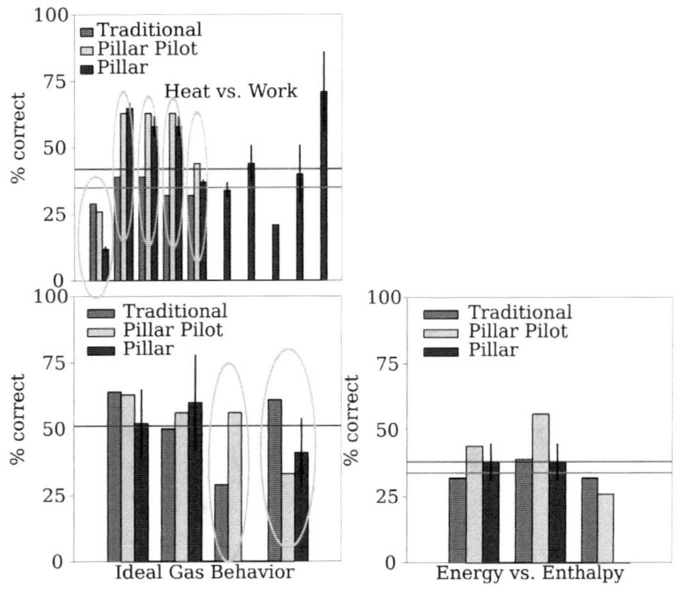

FIGURE 1: *Concept inventory results comparing both traditional and pillar cohorts on thermodynamics concepts. Circled are direct comparisons of concepts that had "significantly" differing scores between cohort groups. As may be noted the pillar cohorts perfomed better on 5 versus 2 concepts.*

relevant pieces of the puzzle. Whenever possible pillars exploit analogies between differing aspects of the discipline both to enhance internalization as well as to decrease instruction time.

The *Foundations of ChE* pillar course combines elements of mass and energy balances, thermodynamics, separations, and product design. This course introduces chemical engineering problem solving techniques from both a (traditional) process-centric viewpoint as well as a product-centric viewpoint. The course spans from theoretical (basic thermodynamics) to applied (separations) allowing a simple route to problem-based learning of difficult theoretical concepts.

The *Thermodynamics* pillar course combines ideas from both pure and multi-component thermodynamics. It introduces molecular insight and the tools (including commercial software) for solving both simple and complex problems in phase and chemical equilibria. The course has a strong focus on multi-scale analysis, for example, covering intermolecular potentials (molecular-scale) to aid students in choosing equations of state for novel materials (macro-scale).

The *Transport Phenomena* pillar course stresses analogies between momentum, mass, and heat transport. Content spans from the molecular origins of transport up through continuum descriptions, as well as macroscopic balances.

The *Reactive Processes* pillar course integrates reactor design, reaction kinetics, and advanced separation processes to allow the comprehensive study of systems ranging from polymerization reactors to enzyme-catalyzed metabolism to (bio-)artificial organs.

The *Dynamics and Modeling* class is the first of a two-part *Systems Engineering* pillar sequence. This course covers dynamical analysis of process systems, process control fundamentals, feedback, basic process modeling, and optimization. The second course in this sequence is the *Design* course which formally combines topics from all other pillars to allow both product and process design.

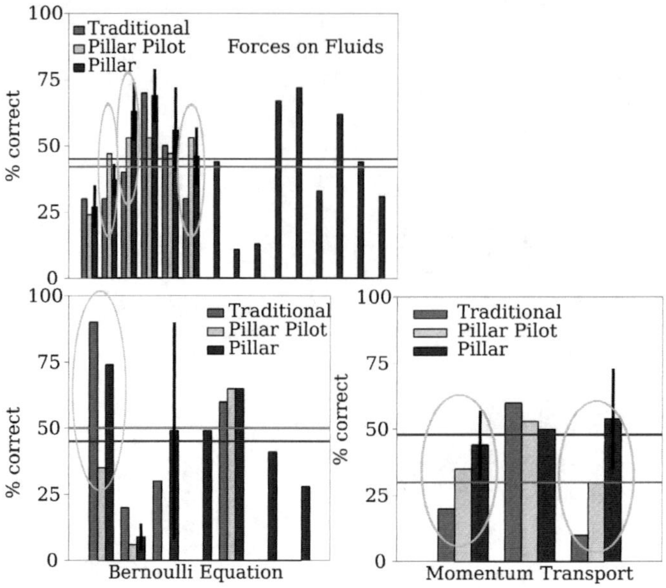

FIGURE 2: *Concept inventory results comparing both traditional and pillar cohorts on transport (fluids) concepts. Circled are direct comparisons of concepts that had "significantly" differing scores between cohort groups. As may be noted the pillar cohorts perfomed better on 5 versus 1 concepts.*

3 Assessment Methodologies and (Some) Results

Our assessment strategy differs from our traditional course assessments due to the inclusion of two new methodologies: concept inventories [9, 10] and concept maps [11–13]. These techniques are aimed at evaluating, respectively, the level of gains in the core areas as well as the level of integration of knowledge attained by the students – a new goal with the revised curriculum.

3.1 Concept Inventories

In order to assess the level of expertise attained in the core areas, and any gains that may be achieved, we are administering concept inventories (CI) [9, 10]. Results for two of our pillar courses are available using existing CIs developed at the Colorado School of Mines (CSM) [9] in thermodynamics (See Figure 1), heat transfer (see Figure 3), and fluid mechanics (see Figure 2). At present, development of additional concept inventories to assess attainment levels in the other four pillar course is the subject of ongoing work, and only the Transport and Thermodynamics pillar course will be discussed here.

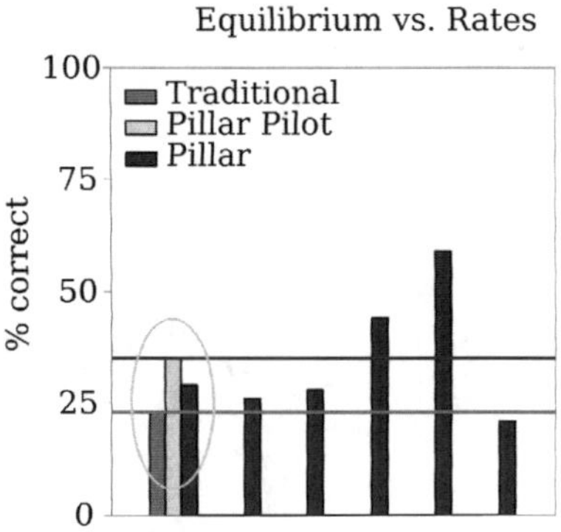

FIGURE 3: *Concept inventory results comparing both traditional and pillar cohorts on transport (heat) concepts. Circled are direct comparisons of concepts that had "significantly" differing scores between cohort groups. As may be noted the pillar cohorts perfomed better on 1 versus 0 concepts.*

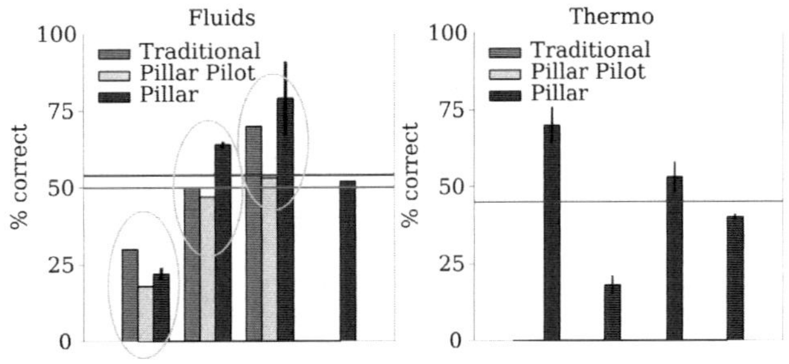

FIGURE 4: *Concept inventory results comparing both traditional and pillar cohorts on cross-cutting (mass vs. energy) concepts. Circled are direct comparisons of concepts that had "significantly" differing scores between cohort groups. As may be noted the pillar cohorts perfomed better on 2 versus 1 concepts.*

Our results are reported in four separate groups – thermodynamics, transport (fluids),transport (heat), and cross-cutting – where the cross-cutting is further subdivided among concepts relating to mass vs. energy and those relating temperature vs. energy. Due to the experimental and evolving nature of the concept inventories used (and the fact that our traditional curriculum cohort group was only available for the earliest versions of the CI's) our results show the raw scores of all related concepts and highlight only significant differences in persistent questions. Furthermore, as an aid to comparison, lines are drawn representing the average (raw) score on the entire suite of related concepts for each cohort group.

As can be seen from the differing classes of concept results reported, we have ample evidence that our change in curriculum has not resulted in a decreased level of mastery of basic (yet challenging) concepts within the discipline. While our results are likely not statistically sound enough to conclude that we have indeed enhanced the learning of our students in these key areas, that was not the aim of the work and we are heartened by the fact that we have "done no harm".

3.2 Concept Maps

In order to assess a student's "systems thinking" or their overall understanding of the topology of the discipline, concept mapping exercises are performed [11–13]. In concept mapping, a student is asked to develop a schematic representation of their understanding of what the discipline of chemical engineering entails. In practice, the students draw a diagram showing a hierarchy of ideas or concepts linked through branches between the sub-concepts, with further links showing interrelationships between inter-branch ideas/concepts, when necessary (i.e., cross-links). A schematic of a concept map is shown in Figure 6.

These maps are "scored" by a (consistent) panel of experts that use a rubric to quantitatively describe the map based on its organization, comprehensiveness, and correctness (see Figure 7). These mapping exercises are performed

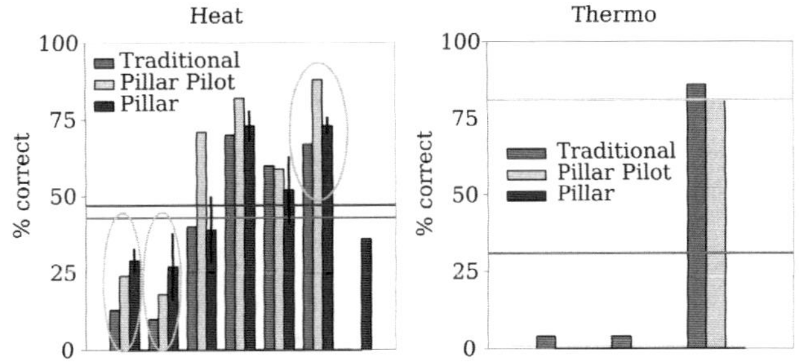

FIGURE 5: *Concept inventory results comparing both traditional and pillar cohorts on cross-cutting (temperature vs. energy) concepts. Circled are direct comparisons of concepts that had "significantly" differing scores between cohort groups. As may be noted the pillar cohorts perfomed better on 3 versus 0 concepts.*

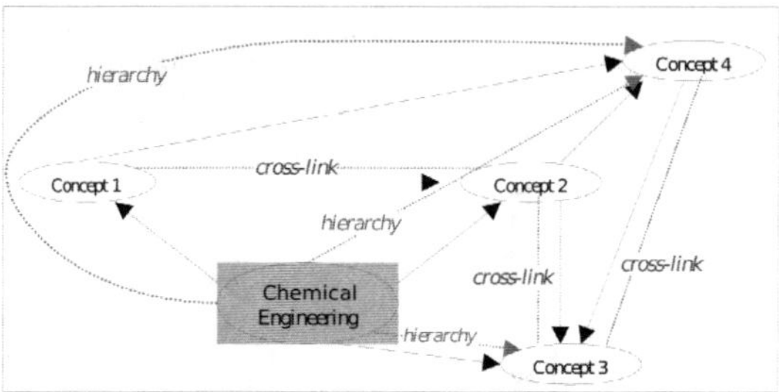

FIGURE 6: *Schematic of a Concept Map. Shown is an example concept map displaying links, branches and cross-links.*

	1	2	3
Comprehensiveness – covering completely/broadly	The map lacks subject definition; the knowledge is very simple and/or limited. Limited breadth of concepts (i.e. minimal coverage of coursework, little or no mention of employment, and/or lifelong learning). The map barely covers some of the qualities of the subject area.	The map has adequate subject definition but knowledge is limited in some areas (i.e., much of the coursework is mentioned but one or two of the main aspects are missing). Map suggests a somewhat narrow understanding of the subject matter.	The map completely defines the subject area. The content lacks no more than one extension area (i.e., most of the relevant extension areas including lifelong learning, employment, people, etc. are mentioned).
Organization – to arrange by systematic planning and united effort	The map is arranged with concepts only linearly connected. There are few (or no) connections within/between the branches. Concepts are not well integrated.	The map has adequate organization with some within/between branch connections. Some, but not complete, integration of branches is apparent. A few feedback loops may exist.	The map is well organized with concept integration and the use of feedback loops. Sophisticated branch structure and connectivity.
Correctness - conforming to or agreeing with fact, logic, or known truth	The map is naïve and contains misconceptions about the subject area; inappropriate words or terms are used. The map documents an inaccurate understanding of certain subject matter.	The map has few subject matter inaccuracies; most links are correct. There may be a few spelling and grammatical errors.	The map integrates concepts properly and reflects an accurate understanding of subject matter meaning little or no misconceptions, spelling/grammatical errors.

FIGURE 7: *Concept Map Scoring Rubric*

for each group of students in the Design Pillar (and preceding transitional capstone design class) and, occasionally, by spring-semester students in earlier levels of the curriculum. In this way we can assess the students' "knowledge integration" not only as a function of curriculum followed, but also temporally within the Pillars curriculum.

Our concept map scores to date are reported in Figure 8 where traditional, transitional, and pillar cohorts are compared (and pillar cohorts include results from sophomores and juniors as well as seniors). The scatter in concept map scores, when reported as means with error bars given as one standard deviation of the scores, makes our assessment inconclusive. Nevertheless, when comparing the median values there is an encouraging trend that not only do pillars students (seniors: 5.83 total score; transitional: 5.50) have a higher score on each measure versus the traditional cohort group (seniors: 4.83), but they are achieving comparable scores to the traditional cohorts earlier in their careers (sophomores: 4.83; juniors: 5.50).

4 Outlook

The Pillars curriculum at the University of Pittsburgh has been lauded by ABET as "a unique integrated approach" that is "in the forefront of ... education". We have no data that our change in curriculum is harmful, in fact, with limited cohort data student performance has seemingly improved on concept inventory tests thus far. Moreover, the median values of the concept map scores show the encouraging trend that our students understand the "big picture" not only better, but also considerably earlier. At the same time, student attitudes have been quite favorable with an across-the-board increase in teaching effectiveness survey scores (not shown) and anecdotal comments such as, "(an) improvement to old style ..." and "... love the pillar system!". Most faculty have embraced the format as it is

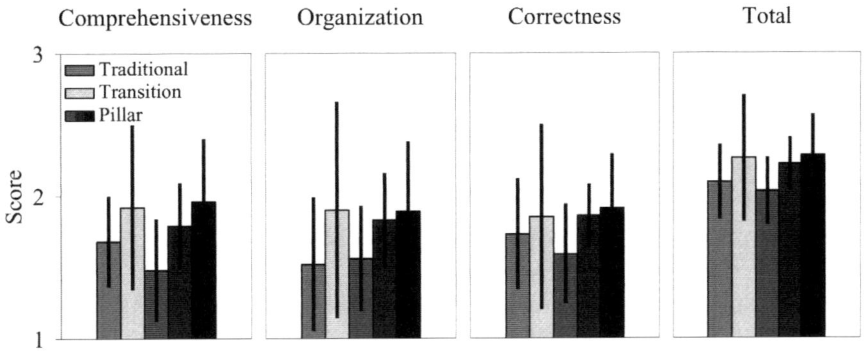

FIGURE 8: *Concept Map Scores. Note that the three shades of blue denote pillars students that have completed sophomore, junior, or seniors years (light to dark), respectively.*

logistically simpler than teaching a comparable number of credit hours in "separate" courses.

As our base of cohort data expands, and the statistics of our assessments (presumably) improves, we hope to port this type of curriculum first to other chemical engineering departments, but ultimately to other disciplines entirely. As a pre-emptive strike against potential issues of textbook availability we have begun an open courseware project, hosted at http://pillars.che.pitt.edu.

Acknowledgment

This material is based upon work supported by the National Science Foundation under Grant No. EEC-0342713.

References

[1] National Science Foundation. NSF engineering education coalitions, 2000. `http://www.eng.nsf.gov/eec/coalitions.htm`.

[2] P. Merton, C. Clark, J. Richardson, and J. Froyd. Engineering curiculla change across the foundation coalition: How they succeeded, what they learned. *Proceedings of the ASEE Annual Conference*, 2001.

[3] JE Froyd and MW Ohland. Integrated engineering curricula. *Journal of Engineering Education*, 94(1):147–164, JAN 2005.

[4] R.L. Canady and M.D. Rettig. *Block Scheduling: A Catalyst for Change in High Schools.* Eye on Education, Inc., Priceton, New Jersey, 1995.

[5] J.J. McCarthy and R.S.Parker. The pillars of chemical engineering: A block scheduled curriculum. *Chemical Engineering Education*, 38(4):292–301, 2004.

[6] R. Glaser. Education and thinking: The role of knowledge. *American Psychologist*, 55:2–21, 1984.

[7] R.L. Canady and eds. M.D. Rettig. *Teaching in the Block: Strategies for Engaging Active Learners.* Eye on Education, Inc., Priceton, New Jersey, 1996.

[8] J. Bordogna, E. Fromm, and E.W. Ernst. Engineering education: Innovation through integration. *Journal of Engineering Education*, January:3 8, 1993.

[9] D. L. Evans. Progress on concept inventory assessment tools. *Frontiers in Education*, 2003.

[10] KE Wage, JR Buck, CHG Wright, and TB Welch. The Signals and Systems Concept Inventory. *IEEE TRANSACTIONS ON EDUCATION*, 48(3):448–461, AUG 2005.

[11] JR McClure, B Sonak, and HK Suen. Concept map assessment of classroom learning: Reliability, validity, and logistical practicality. *JOURNAL OF RESEARCH IN SCIENCE TEACHING*, 36(4):475–492, APR 1999.

[12] M Besterfield-Sacre, J Gerchak, MR Lyons, LJ Shuman, and H Wolfe. Scoring concept maps: An integrated rubric for assessing engineering education. *Journal of Engineering Education*, 93(2):105–115, APR 2004.

[13] J. Gerchak, M. Besterfield-Sacre, L.J. Shuman, and H. Wolfe. Using concept maps for evaluating program objectives. *Frontiers in Education*, 2003.

A Spiral Curriculum for Chemical Engineering

Vinay K. Gupta, Babu Joseph, Norma Alcantar, Ryan Toomey and Aydin Sunol,*
Chemical & Biomedical Engineering, University of South Florida, 4202 E Fowler Avenue, ENB 118, Tampa, FL 33620

1. Introduction

The broad goal of our work is to transform the educational experience of undergraduate students in chemical engineering. Towards this end, we are developing and implementing a "multidimensional spiral curriculum" by integrating concepts from core courses at the sophomore, junior, and senior level. In the first step described below, our implementation focuses on students who transfer from two-year community colleges into a four-year baccalaureate program at the University of South Florida (USF).

2. Spiral versus sequential curriculum: critical considerations

The traditional curriculum that is offered in most engineering departments is sequential in nature and students cover core topics in their major discipline through a series of courses. For example, in Chemical Engineering departments across the US a student starts with basic concepts such as material and energy flows in chemical processes (Figure 1).

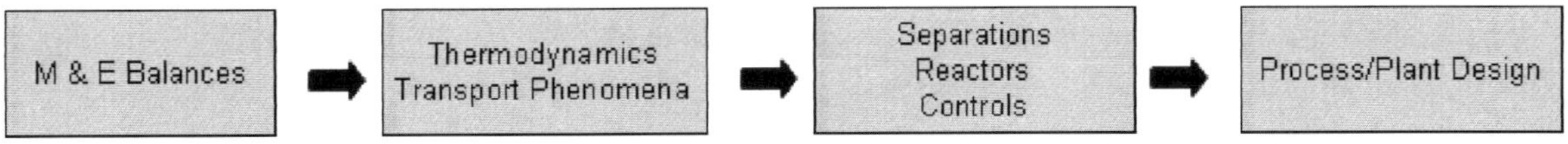

Figure 1. Traditional Sequence of Core Topics in a Chemical Engineering Curriculum

Typically, this is followed by thermodynamics and transport of mass, momentum, and heat courses. Progression through the final years of a degree program involves courses on chemical separations, reaction engineering, process control, and a capstone design course. These core courses are supplemented with various technical electives that differ from one program to the other or depend on the professional track that a student will follow. The sequential arrangement separates the core curriculum into smaller units, expose students to each unit in succession, and require mastery of each unit before stepping on to the next unit.

For a student in a four-year program, the traditional sequential curriculum has both advantages and disadvantages. From an institutional perspective the sequential curriculum is easier to teach. The modularity of a sequential curriculum provides structure to the course offerings for an engineering discipline and defines boundaries to the content of courses for a degree program. Proponents of sequential curriculum argue that this approach is better from a student perspective as well since the concept of "Mastery Learning" wherein one ensures that students have mastered each increment of a subject in a hierarchical sequence before going

* Author to who correspondence should be addressed: vkgupta@eng.usf.edu

on to the next is a superior learning approach. On the other hand, in each course within a sequential curriculum a final examination is the primary mechanism to test the competency of the student in the topic that is studied and satisfactory performance allows a student to advance to the next stage of study. Thus, the sole motivation for a student is to work for a passing grade in one course and progress to the next course in sequence. Typically, little consideration is given on how the knowledge and skills in one course will be used in subsequent courses or its contribution to the whole curriculum. Implicitly, this integration between concepts is left to students and is expected to happen automatically in the capstone design course. Furthermore, even though there are many unifying and common concepts between courses, there is no systematic emphasis on these linkages unless an individual instructor focuses on this unification at his or her own discretion.

Regardless of these merits or demerits, it is clear that the linear sequence approach to the curriculum fits best to programs where students can begin in a specific major discipline as freshmen and proceed through the series of courses towards graduation. For students that transfer from two year community colleges, the disadvantages of the sequential offerings become exacerbated because they cannot benefit from the gradual spacing of courses.

The spiral curriculum is an alternative model to the sequential curriculum and was first proposed in 1960 by an educational theorist named Jerome Bruner [1]. Here, a set of interlinked and basic concepts are presented briefly and simply to students on a first traverse around a spiral path (Figure 2). As a learner completes a traverse, the thread is then continued at a higher level around the fundamental ideas of the spiral in greater detail, breadth, and sophistication. Thus, the spiral focuses on introducing higher cognitive content with progress along the spiral path. The model of a spiral curriculum has been frequently applied in K-12 education of science and mathematics topics [2], where it is often referred to as *incremental learning*. In higher education, the spiral curriculum has only recently gained an entry with applications in the study of nursing, medicine, and optometry [2-5], design of a course for non-computer science majors at Michigan State University[6], and the sophomore year for chemical engineering students at Worcester Polytechnic Institute[7].

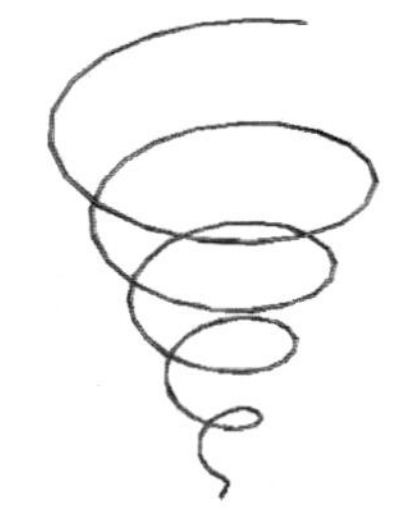

Figure 2

3. Multi-dimensional spiral curriculum in chemical engineering

Our goal is to address student learning by undertaking a curriculum transformation in Chemical Engineering such as that depicted in Figure 3a where all the core concepts of material & energy balances, thermodynamics, transport processes, reaction engineering, controls, and design are simultaneously present in a spiral curriculum. Figure 3b shows a schematic of the **multi-dimensional spiral curriculum** that we propose to implement for transfer students using three inter-locking spiral paths. Along one direction of this curriculum the students will progress from the conceptually simple steady state systems and equilibrium units to the more complex task of large scale systems integration while in the second direction the students will learn the core technical concepts for analysis and design that are essential for a professional chemical engineer.

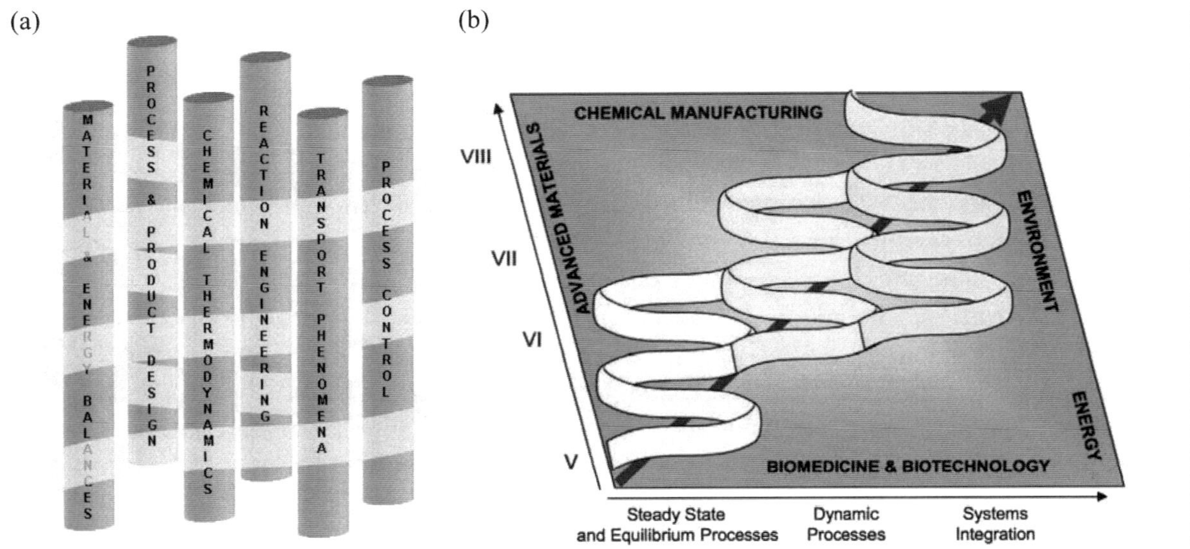

Figure 3. Schematic illustration of (a) spiral model adapted for core curriculum in chemical engineering, and (b) a multi-dimensional spiral curriculum.

For instance, the students may begin with a material balance in non-reactive systems and upon revisiting the concept a second time may learn how to incorporate reactions and on the third time may learn how to go from macroscopic to differential balances. Concurrently, students may begin with phase equilibrium for a single substance first that is necessary for material balances, then learn about phase equilibrium in multi-component ideal mixtures on the second visit, and then learn about non-ideality, which requires a higher level of sophistication. Thus, a spiral curriculum puts into practice the philosophy that "Learning is an organized process with acquisition of new knowledge requiring prior core knowledge"[8].

In this spiral approach, *links between different traverses* of the spiral are emphasized which allows students to reinforce the prior concepts and begin grasping the application to new knowledge. For example, when students learn heat exchanger design and heat transfer, the link to first solving a mass balance is stressed. *Vertical integration* between learning loops is achieved by choosing themes that are central to the overall outcome that is desired. For example, if the development of skills to design a chemical product or process (e.g., energy requirements of an automobile, drug delivery system, miniaturized chemical assay system) is a desired outcome then the spiral of core topics is sequenced around one or more illustrative chemical systems (e.g., a fuel cell, drug loaded nanoparticles, microfluidic device, respectively) in increasing level of sophistication as each loop of the spiral is traversed.

A steady and continuous **deepening of knowledge** is incorporated in the spiral curricular approach as students move from simple to complex concepts. **Organization and**

integration are implemented in the spiral approach as the complex relationships between different engineering topics can be correlated and sequenced. This feature is especially relevant in current times where a multitude of engineering educators are focusing on integrated curricula. Our implementation also takes advantage of **reinforcement based learning,** where a spiral approach allows students to be exposed continuously to previously learned topics.

4. Results and Discussion of First Cycle of Implementation (2006-2008)

Implementation of the curriculum and enrolment of transfer students in the new curriculum began during 2006-2007 academic year. From a common pool of students that typically begin their studies in chemical engineering at USF, three groups of students were identified. **Group 1** included students who had not taken a pre-requisite course in engineering thermodynamics. This group composed largely of students who began at USF in the College of Engineering in a traditional sequential curriculum. It also included a few transfer students who did not consent to be part of the curriculum reform research. **Group 2** was composed of students who had taken a pre-requisite course in engineering thermodynamics either at USF or an equivalent course elsewhere. Students in **Group 3** were transfer students who consented to be part of the new curriculum. The range of GPA of the students in the Group 3 was similar to the students that were in Groups 1 and 2.

The students participating in the multi-dimensional spiral curriculum were assessed in several different ways. Figure 4 indicates the different paths that the students in the three groups took during the first implementation. It is clear that intermediate stages such as the end of year 1 exist where students have cumulatively experienced course topics that are quite similar in range though by different instructors. Therefore, common quantitative exams in the courses indicated by a green border were used as a tool to compare student proficiency and grasp of concepts in the different groups. Additionally, there are several other points of intersection where the performance of students can be assessed alongside one another. We chose three laboratory courses because they offer an opportunity to test how well students can build on concepts from the classroom courses and it requires students to engage in hands-on activities in teams. They have to write reports, give oral presentations, and also take traditional quizzes. Student performance was also tracked in classroom courses such as Numerical Methods, Materials Science, Engineering of Biological Systems, and Capstone Design. These courses were not part of the transformed curriculum and therefore, offered an opportunity to assess students without ambiguities from different instructors or class size. Finally, the students were asked to complete exit surveys and participate in an exit interview at the time of graduation for assessing their perceptions of the curriculum.

As an illustrative example, Figure 5 shows the plot summarizing the comparison of student performance in the three common laboratory courses. In the plot, the maximum, minimum, and average grade points for the letter grades are shown for students in Group 3 separately from all other students. Even though a direct cause and effect relation is difficult to establish because of differences in class size for the spiral and non-spiral courses, the plot in Figure 5 indicates that range of grades and the average for the students who participated in the novel spiral curriculum was higher. Similar trend was found in the other common courses suggesting that curriculum reform and implementation can be lead to valuable gains.

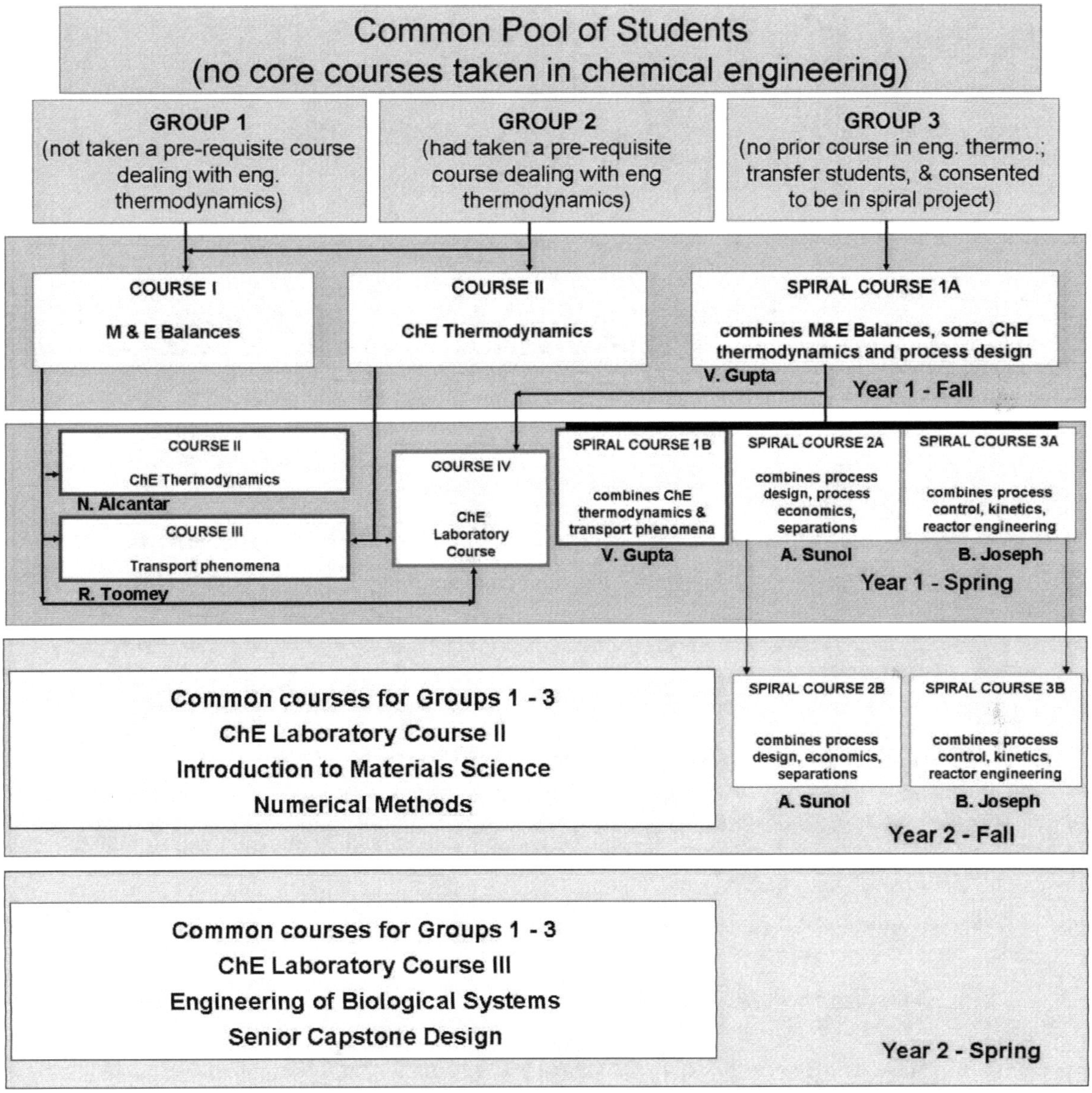

Figure 4

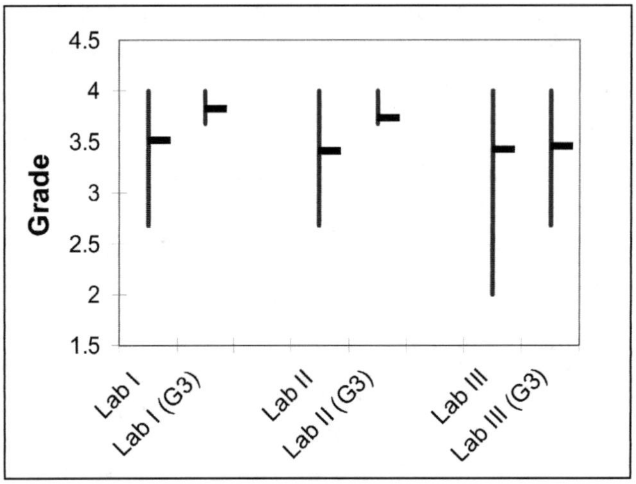

Figure 5

REFERENCES CITED

[1] J.S. Bruner, *The Process of Education*, Harvard University Press, Cambridge, MA, (1960).

[2] T.J. Dowding, *Educational Technology* **33**,(1993), 18-28.

[3] Beattie, *Making a curriculum work*, in: M. Jolley, P. Allan (Eds.), Curriculum Issues in Nurse Education, Croom Helm, London, **1986**.

[4] R.M. Harden, M.H. Davis, J.R. Crosby, *Medical Education* **31**,(1997), 264-271.

[5] J.J. Kabara, *Journal of Medical Education* **47**,(1972), 314-316.

[6] M. Urban-Lurain, *Fluency with information technology: The computer science perspective*, in: Y. Zhao (Ed.), What should teachers know about technology: Perspectives and practices, Information Age Publishing, Greenwich, CT, **2003**.

[7] W.M. Clark, D. DiBiasio, A.G. Dixon, *Chemical Engineering Education* **34**,(2000), 222-233.

[8] E.D. Hirsch, *Educational Leadership* **50**,(1993), 23-25.

History of the Summer Schools for Chemical Engineering Teachers (Faculty)

Deran Hanesian
Master Teacher and Professor of Chemical Engineering
New Jersey Institute of Technology
Newark, NJ 07102

Ralph A. Buonopane
Associate Professor Emeritus of Chemical Engineering
Northeastern University
Boston, MA 02115

Angelo J. Perna
Master Teacher and Professor of Chemical Engineering and Environmental Engineering
New Jersey Institute of Technology
Newark, NJ 07102

The Summer Schools for Chemical Engineering Teachers (Faculty) are a unique educational experience developed as a means of transferring the latest educational methods and technical information by experienced faculty to new, younger faculty, in essence, a mentoring process to improve teaching. Its impetus came from the Summer Schools for Engineering Teachers program developed by the Society for the Promotion of Engineering Education (S. P. E. E.). However, it all started with engineering.

The Summer Schools for Chemical Engineering Teachers are a unique event designed to have a major impact on the profession by:

1. Disseminating information to improve teaching and course content.

2. Furthering the development of individual chemical engineering educators.

3. Introducing new applications of chemical engineering.

4. Building bridges between industry and the academic world.

These purposes are accomplished through the formal and informal interaction of representatives from departments of chemical engineering with each other and with invited industrial representatives. The academic attendees to the Summer School were at the discretion of individual Chemical Engineering Department Heads. How did it all begin?

The concept of Summer Schools for Engineering Faculty began to germinate in 1925 by the Society for the Promotion of Engineering Education. William E. Wickenden, a former engineering teacher and later acting Vice President, American Telephone and Telegraph, was engaged in investigating and coordinating the educational activities of the Bell System. He became the Director of the S. P. E. E, Board of Investigation and Coordination and was assisted by Professor H. P. Hammond, as the Associate Director. Studies of Engineering Education commenced in 1923 with the financial support of the Carnegie Foundation of $108,000 for a three year study.

In 1925, Professor Wickenden was aware of a recent survey of technical education in Europe, which showed that there were possibilities of direct borrowing from the Europeans. He went to England and was very impressed with what he observed and felt that this concept could be transplanted to America and would *"take root in American soil"*.

Two professional societies led to the Summer Schools for Chemical Engineering Faculty. One was, "The Society for the Promotion of Engineering Education", founded in 1893 by 70 men who met and discussed educational papers. These men were very interested in enhancing the important issues of engineering education, in general, for all engineering disciplines.

Similarly, the other was, the "American Institute of Chemical Engineers" founded in 1908 by 40 men with the purpose of enhancing the chemical engineering profession. The founding leaders of the American Institute of Chemical Engineers knew that their hope for acceptance as a distinct profession was through education. This controversy lasted 22 years until 1930, when the chemical engineering professionals proved to be the leaders, among all professional societies, in accreditation of university programs.

The Summer School for Engineering Teachers program in the United States was inaugurated by two sessions in 1927. The sessions were under the jurisdiction of the Board of Investigation and Coordination of SPEE, but Local Arrangements Directors were named at each university. It was proposed that the Summer Schools be held immediately after the 35th Annual Meeting of the Society at the University of Maine, Orono, Maine, which was scheduled for June 27-30, 1927. The duration of the summer school would be two and one-half to three weeks long. The first school was to be held at Cornell University, Ithaca, New York with Dean Dexter S. Kimball in charge and the second at the University of Wisconsin at Madison, Wisconsin with Dean F. E. Turneaure in charge. The subject of the summer schools was to center around a particular subject with the attendees being the younger teachers. *"The number of each session will be limited to an efficient working group. There will be a staff of outstanding teachers and the work will be conducted on an intimate basis; much of it by the seminar method"*. The subject selected for these first two sessions was Mechanics, both pure and applied because it was felt that this subject was basically important for all branches of engineering. It was felt that the subject of mechanics would provide an ideal test case for this trial year. The subject is *"taught in every curriculum; it occupies the pivotal position between the general sciences and the more specific technical subjects; it runs a constant risk of becoming conventionalized; and it is capable of becoming notably serviceable to subjects which follow if taught with vision and vigor"*

"The bringing of groups of the ablest and most promising younger teachers into contact with the leaders having different points of view and different methods of presentation with adequate opportunity for free discussion promises a forward step in methods of teaching, in developing and holding the interest of the student, and in putting new life and inspiration into this department of engineering training".. Hence, the first step had been taken for the improvement of teaching methods and the training of engineering teaching personnel.

As a result of these summer schools, committees with a secretary were formed to act for the entire group until the next summer school for teachers of the same subject to maintain continuity.

Following the successful experience with the first two Summer Schools, others followed under the leadership of the S. P. E. E. These were:

1928	Physics	Massachusetts Inst. of Tech.
1928	Electrical Eng,	University of Pittsburgh and Westinghouse Electric and Manufacturing Company
1929	Mechanical Engineering	Purdue University
1930	Civil Engineering	Yale University
1930	Drawing and Des. Geometry	Carnegie Inst. Of Tech.

1931	Chemical Engineering	University of Michigan
1931	Mathematics	University of Minnesota
1932	Economics	Stevens Inst. Of Tech.
1932	English	Ohio State University
1933	Mining and Metallurgical Engineering	University of Wisconsin

This completed the cycle of sessions that was begun when the Summer School was established in 1927. SPEE gave the responsibility of future Summer Schools to the professional societies. The professional societies did not conduct Summer Schools regularly after 1933 but reverted to professional conferences at the SPEE meetings.

Committee 22 for Chemical Engineering had been formed in about 1927 and then in 1938 it became the Chemical Engineering Division (CHED). However, the first Chemical Engineering Summer School was by a petition to SPEE from the Board of Directors of the American Institute of Chemical Engineers (AIChE). This was the beginning of a long term interaction between CHED and AIChE.

The Summer Schools for Chemical Engineering Teachers have listed the following objectives over the years for the fourteen (1931-2007) Summer Schools for Chemical Engineering Teachers. The importance of the Chemical Engineering Division, ASEE Summer School for Chemical Engineering Faculty is:

- it is essential to timely progress in chemical engineering education
- it is unique to the discipline of chemical engineering and involving 100 educators
- it is offered only every five years
- its emphasis on new faculty hired within the last five years
- it is targeted to provide innovative and effective teaching methods
- it provides educational classroom materials on latest technologies and research
- its participation by exceptional mid-career and senior faculty as mentors
- that all chemical engineering departments are invited to send participants
- the support and participation by major corporations and federal agencies to accomplish through the formal and informal interaction of representatives from departments of chemical engineering with each other and with invited industrial and government representatives.
- its results and materials will be widely distributed by the latest available methods of improved, advanced communication.

The dates and locations of these events were:

1st June 22–July 9, 1931 University of Michigan, Ann Arbor, Michigan

Alfred H. White, Professor of Chemical Engineering at the University of Michigan, was Director and W. L. McCabe, Assistant Professor of Chemical Engineering was Secretary.

2nd June 23–June 30, 1939 Pennsylvania State University, State College, Pennsylvania

R. C. Kintner of the Armour Institute of Technology was the Director and F. C. Whitmore and D. S. Cryder both of Penn State were in charge of local arrangements.

3rd August 30–September 4, 1948 The University of Wisconsin, Madison, Wisconsin

The Executive Committee of CHED of ASEE for 1947-48 undertook the responsibility of organizing and operating the 3rd Summer School. The committee consisted of

R. A. Ragatz, University of Wisconsin, Chairman
E. M. Schoenborn, North Carolina State College, Past Chairman
W. A. Koehler, University of West Virginia, Vice Chairman
J. D. Lindsay, A & M College of Texas, Vice Chairman
F. M. Taylor, Tulane University, Secretary
J. H. Koffolt, Ohio State University, Member of ASEE General Council

Theme **"What Should We Teach And How Can We Teach It"**

4th June 27–July 2, 1955 Pennsylvania State University, State College, Pennsylvania

Kenneth A. Kobe, University of Texas, General Chairman and F. L. Carnahan of Penn State served as Local Arrangements Chairman.

Theme: **"What's Ahead in Chemical Engineering Education?"**

5th August 20–August 25, 1962 University of Colorado, Boulder, Colorado

Lloyd Berg, Montana State College, Director and General Chairman
B. E, Lauer of the University of Colorado was the Chairman of Local Arrangements.

Theme: **"Advances in Chemical Engineering Education"**

6th June 20–June 24, 1967 Michigan State University, East Lansing, Michigan

Lloyd Berg, Montana State University, General Chairman
D. K. Anderson of Michigan State University was Chairman, Facilities and Housing.

Theme: **"Dynamic Objectives of Chemical Engineering Education"**

7th August 13–August 18, 1972 University of Colorado, Boulder, Colorado

L. Bryce Andersen of Newark College of Engineering was the Chairman of a Planning Committee

8th July 31–August 5, 1977 Snowmass, Colorado

Co-chairmen: C. Judson King and Michael C. Williams of the University of California, Berkeley.
J. Peter Clark, Virginia Polytechnic Institute and State University, Local Arrangements

Theme: **"New Applications of Chemical Engineering".**

9th August 1–August 6, 1982 University of California, Santa Barbara, California

Co-chairmen: T. W. Fraser Russell and Stanley T. Sandler, University of Delaware at Newark, Delaware.
D. E. Seaborg, University of California, at Santa Barbara, Local Arrangements

Theme: **"Chemical Engineers Need to Have an Impact on Society in a Broader Sens"**.

10th August 9–August 15, 1987 Southeastern Massachusetts University, North Dartmouth, Massachusetts

Co-Chairman: Glenn L. Schrader and Maurice A. Larson, Iowa State University in Ames, Iowa
L. Bryce Anderson, Dean of Engineering at Southeastern Massachusetts University and Stanley M. Barnett of the University of Rhode Island, Local Arrangements

Theme: **"The Revitalization of the Chemical Engineering Curriculum"**

11th August 9–August 15, 1992 Montana State University, Bozeman, Montana

Ralph Buonopane, Northeastern University, Chairman
John T. Sears and Ron Larson, Montana State University, Local Arrangements

Theme: **"Frontiers in Chemical Engineering Education—Curriculum and Needs for the Next Century"**

12th August 10–August 16, 1997 Snowbird, Utah

Bruce Finlayson, University of Washington, Director
Lamont Tyler and Edward Trujillo, University of Utah, Local Arrangements

13th July 27–August 1, 2002 University of Colorado, Boulder, Colorado

Co-chairmen: Scott Fogler and Michael Cutlip

14th July 28 – August 2, 2007 Washington State University, Pullman, Washington

Co-chairmen: Kirk Schulz, Mississippi State University and Steve LeBlanc, University of Toledo
Richard Zollars, Washington State University and Douglas K. Ludlow, University of Missouri, Rolla

Each of the Summer Schools, generally, has a memorable event associated with it that remains in the attendees' mind over the years.

Despite the natural changes that have occurred in Summer Schools over the years, Hammond, in his final report on the Summer Schools from 1927-1933, correctly stated, *"throughout the undertaking, however, the aim has been the attainment of one principal purpose: the improvement of classroom teaching".*

Over the seventy seven years of the Chemical Engineering Summer Schools, the goals have remained consistent in promoting excellence in teaching, introducing the new areas of chemical engineering, development of camaraderie among chemical engineering faculty, and furthering the development of the younger chemical engineering faculty by mid-career and senior Faculty members,

621b a Director's Reflections on the CHED Summer School

Judson King[1], T. W. Fraser Russell[2], Stanley I. Sandler[2], Ralph Buonopane[3], Kirk H. Schulz[4], and Steven E. LeBlanc[5]. (1) University of California, Berkeley, South Hall Annex #4650, Berkeley, CA 94720-4650, (2) Chemical Engineering, University of Delaware, 150 Academy St, Colburn lab, Newark, DE 19716, (3) Chemical Engineering, Northeastern University, Boston, MA 02115, (4) Vice President for Research and Economic Development, Mississippi State University, 617 Allen Hall, Mississippi State, MS 39762, (5) Chemical & Environmental Engineering, University of Toledo, Toledo, OH 43606

Past directors of the Chemical Engineering Summer School offer their perspectives on the event. Speakers include:

1977 Snowmass, CO, Judson King, U of Calif, Berkeley 1982 Santa Barbara, CA, T.W.F.Russell, S.I. Sandler, U. of Delaware

1992 Bozeman, MT, Ralph Buonopane, Northeastern U. 2007 Pullman, WA, S.LeBlanc, U. of Toledo and K. Schulz, Miss. State U.

Participants in the Summer School will also comment on the experience and benefits.

626a Some Thoughts on Chemical Engineering Design Education

Jeffrey J. Siirola, Eastman Research Division, Eastman Chemical Company, PO Box 1972, 200 South Wilcox Drive, Kingsport, TN 37662-5150

Every engineering curriculum requires a culminating capstone design experience. As the chemical engineering discipline evolves, what aspects of design should be taught? Formulation, synthesis, analysis, evaluation, optimization? Process or product? Batch or continuous? Device or specialty or commodity? Chemical or biological or electronic? Creative or routine? At what scale? Involving what tools? This presentation will outline some basics and innovations that may help students better meet future industrial chemical engineering design challenges.

626b The Product/Process Design Dichotomy

Warren D. Seider, Chemical and Biomolecular Engineering, University of Pennsylvania, Philadelphia, PA 19104, J. D. Seader, University of Utah, Dept. of Chemical Engineering, University of Utah, Salt Lake City, UT 84112, Daniel R. Lewin, Technion, PSE Research Group, Chemical Engineering, Technion, Haifa, 32000, Israel, and Soemantri Widagdo, Display and Graphics Business, 3M Company, 3M Center, Building 236-1B-17, St. Paul, MN 55144.

This presentation addresses the interface between process and product design education. Three classes of chemical products are considered: (1) basic chemicals, (2) industrial chemicals, and (3) configured consumer chemical products. Beginning with an innovation map, to relate new technologies to consumer needs, the stages of product design are considered with emphasis on the Stage-Gate Product Development Process. Emphasis is placed upon the first two stages, the concept and feasibility stages – with reference to several case studies.

626c Difficulties in Teaching Chemical Product Design

Edward L. Cussler, Chemical Engineering and Materials Science, University of Minnesota, 151 Amundson Hall, 421 Washington Ave. SE, Minneapolis, MN 55455

More chemical engineering graduates are now hired by product-oriented companies than in process-dominated industries. To prepare these students, we can teach chemical product design for three types of products: devices, molecules, and microstructures. Chemical devices, exemplified by home oxygen supply and by domestic ultrafiltration of water, are designed with traditional unit operations. Molecules, like the salmon pigment astaxanthin and the prostate cancer drug Zoladex, also depend for their manufacture on unit operations. But microstructured products, like sunscreen and breakfast cereal, require inventing new tools for development. This paper will describe efforts to develop these tools.

The Value of Batch Process Design in a Chemical Engineering Education

by Jack Vinson, PhD[+]
Product Manager, Batch Process Development
Aspen Technology

Abstract:

Process design for the batch industries, particularly pharmaceutical process design, relies heavily on collaboration across technical functions and across time. Whereas traditional chemical development can occur in just a few months, pharmaceutical development spans many years and has many requirements beyond providing a manufacturable process. This has interesting applications in the education of tomorrow's chemical engineers, particularly with respect to providing guidance on how traditional chemical engineering activities are aligned closely with the larger efforts of the business.

1 Background

A batch process is any manufacturing process that runs in short time bursts, where the quantity or scale of manufacture does not justify continuous operation. Nearly all pharmaceutical production is done in batches, for both the active pharmaceutical ingredient (the chemical) and the drug product (the pill). Specialty chemicals, by their nature of being "specialty," are commonly produced in batches. These might be high-performance materials, new chemicals looking for a market, some agricultural chemicals, inks and paints, and many others. Biochemical processes, depending on their scale, are run in batches or semi-continuously. Commercial food production is often done in batches, though there are aspects that appear semi-continuous. Alcohol production is batch wise—even mass produced beers are fermented in large batches.

Chemical engineers learn this from the outset of their education; the cut-off between batch and continuous production is one of the first things discussed in chemical engineering 101 courses, along with the basics of doing material and energy balances. After this discussion, the chemical engineering education—at least when it comes to production-scale—is focused almost entirely on continuous production environments. Given that the vast quantity of chemical production is done via continuous production, this makes sense.

However, chemical engineers have been moving into non-traditional markets, where batch production dominates. A great value of chemical engineering concepts is that the basics apply in non-traditional markets just as they apply in traditional chemicals and petroleum endeavors. While engineering schools are adding deeper study of batch processes and biotechnology, there are aspects of the larger business environment that should be included in the education of future engineers. These areas of focus include collaboration, the time it takes to develop manufacturing processes, and regulatory hurdles.

2 Batch Process Development

Within the business, the general goal of batch process design is the same as any other process design work: devise a process that is safe, documented, repeatable, efficient, and meets regulatory requirements. The work that goes into a process design happens at various stages throughout the business cycle of bringing a new product to market. Figure 1 gives a view of where batch process design fits into the larger business flow for process development. None of these boxes are labeled "design" because it happens throughout the development workflow.

[+] I want to thank my colleagues, past and present for inspiring and adding to the thoughts and ideas presented here.

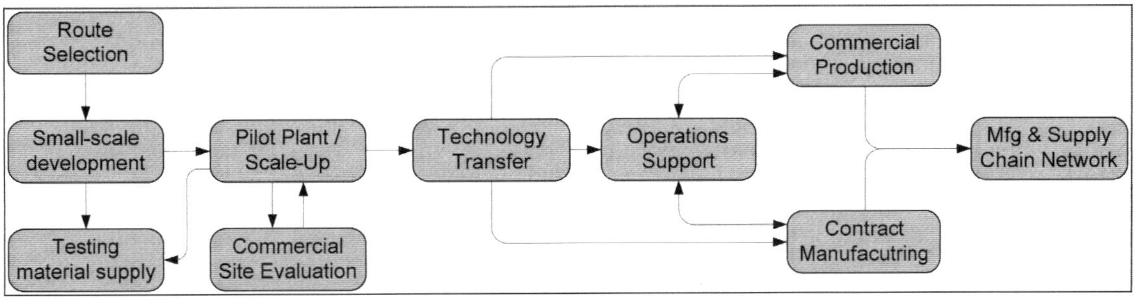

Figure 1: A view of batch process development activities

At the earliest stage, a new product arrives through the research organization (or via contract from partner organizations). The first stage of development, **route selection**, is to design the basic set of chemical/biochemical transformations to make the product at a reasonable scale from the commonly available raw materials. Along with the key bio/chemical expertise, the process of route selection seeks input from process development and production engineers for the kinds of operations that are most easily managed in the available equipment. In addition, the procurement organization will provide estimated costs for raw materials. Even at this early stage, organizations will begin to consider where they intend to do long-term manufacture of the product, and they will start negotiating with suppliers for price on raw materials or key production steps. Figure 2 is an example of a route view of the process.

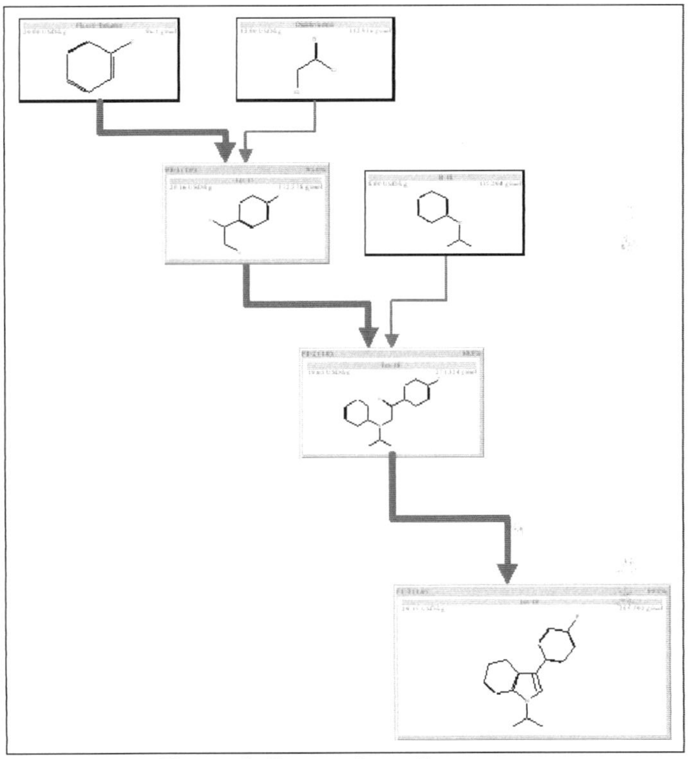

Figure 2: Route view of a process

With a basic route selected, the process goes through a series of **scale-up** cycles from laboratory scale to kilo labs to a pilot plant and eventually into manufacturing. At each stage, more and more detail is added to the process design and to the know-how for this process. A process that may have been described as a simple sequence of reactions on a single sheet of paper grows to become a recipe in the lab (such as in Table 1), and it eventually expands to a full set of operating instructions that can be several hundred pages long. Along with the scale-up activities, the organization must supply materials for testing, formulation studies, packaging trials, market trials, clinical trials, and other needs.

At some point during the process, the business must decide how and where to make the product, the **site evaluation**. The decision depends on how well the process fits into the available plants from both a processing perspective and the anticipated load on the facility. Process fit involves exploring whether the equipment is sufficient to achieve the process conditions of the product. The load question involves demand forecasts for existing and future products that are made in the plant. To handle the highly uncertain nature of new product demand, some companies set aside "launch" facilities, where they plan to manufacture a new product for its first several years and then shift to another facility for long-term manufacture. If no facility will fit the process, decisions may arise about expansion and construction of greenfield sites. Alternatively, a company may decide to use a contract manufacturer to produce some or all of their product. This, of course, raises another set of issues related to managing the process and guaranteeing quality, but contract manufacturing is on the rise as companies focus on their core skills of innovation and marketing.

Table 1: A simple text recipe

Aspirin R&D Process

1. Reaction
   - 1.1. Charge Erlenmeyer Flask, 250 ml with 0.003 kg of SALICYLIC-ACID. Dissolve the following components: 100% of SALICYLIC-ACID.
   - 1.2. Charge Erlenmeyer Flask, 250 ml with 0.006 liter of ACETIC-ANHYDRIDE.
   - 1.3. Charge Erlenmeyer Flask, 250 ml with 0.001 liter of 85%H3PO4.
   - 1.4. React in unit Erlenmeyer Flask, 250 ml via Aspirin Synthesis. The final temperature of the batch is 75 C.

----

2. Crystallize
   - 2.1. Charge Erlenmeyer Flask, 250 ml with 0.02 liter of WATER.
   - 2.2. Crystallize the batch in unit Erlenmeyer Flask, 250 ml. The following components are separated in the crystal phase: 100% of ACETYLSALICYLIC-ACID. Ramp the temperature to 25 C.

----

3. Filtration
   - 3.1. Filter the batch from unit Erlenmeyer Flask, 250 ml in filter Filter Flask, 250 ml.
   - 3.2. Wash the cake in unit Filter Flask, 250 ml. For each wash, use 0.0025 liter of WATER. Spent Wash Stream: The stream belongs to category: Solvent Recovery.

----

4. Drying
   - 4.1. Transfer contents of unit Filter Flask, 250 ml to Evaporating Dish.
   - 4.2. Dry the batch in unit Evaporating Dish. The drying time is 30 min. The drying temperature is 100 C. Evaporated Solvent: The stream belongs to category: Solvent Recovery.

At the right time, the development activities shift from scale-up and tweaking the process to transferring the process, knowledge, and responsibility to the manufacturing location. During **technology transfer**, there are many coordinating activities and requirements for a successful transfer, such as validated trial runs that show the manufacturing site can produce within specifications. This is a key requirement in regulated industries, but it is important for many high-value products, as slight deviations can have a detrimental impact on the final quality or performance of the product in the marketplace.

Once settled into manufacturing, organizations have the opportunity (and obligation) to continue monitoring the performance of the process and the quality of the final product. While it is ideal to have the best process in the plant from the outset, that is difficult to achieve in reality, particularly since new technology and new regulatory oversight continue to change expectations. Even without these changes, the manufacturing site is subject to its own changes in product mix and demands from the market, so the manufacturing process must be updated continually and improved to fit the current circumstances.

Depending on the business, this sequence of steps may happen all together over the course of a few months (specialty chemicals), or it may happen over several years (pharmaceuticals), or somewhere in between. On the shorter time scales, the development activity may go right from a new product to designing a process for manufacturing scale, eliminating or reducing a number of activities in Figure 1.

While this description focuses on processes that include at least one chemical or biochemical transformation, the general concepts apply to formulation development: creation of the full-scale process that meets the safety and quality requirements, and within the appropriate time-frames.

3 How is batch different from continuous?

After understanding the basics behind batch process development, the key elements that are unique to the batch environment begin to surface. Some of them have to do with designing and managing the process itself. Other differences are associated with how process development fits into the larger framework of the business and its goals.

On the process design side, the main difference is on the focus of the design work: the product itself. The typical approach is to design the transformations, find appropriate solvents and anti-solvents for processing, and use fairly standard unit operations to process the materials. At the **pilot plant/scale-up** stage (Figure 1), the detailed development of a step looks very similar to traditional conceptual design for chemical engineering, which focuses on designing and optimizing the unit operations, once the transformations have been decided.

Recycle strategies in the batch industries are different. There is no continuous recycle with purge and make-up streams. At best, manufacturing facilities have solvent (and other) recovery systems to provide fresh, to-specification materials for use in subsequent batches, or in other processes. Design for recycle in batch operations focuses on ensuring that impurities can be removed in the recovery operations, so on-specification materials feed back into the next batches.

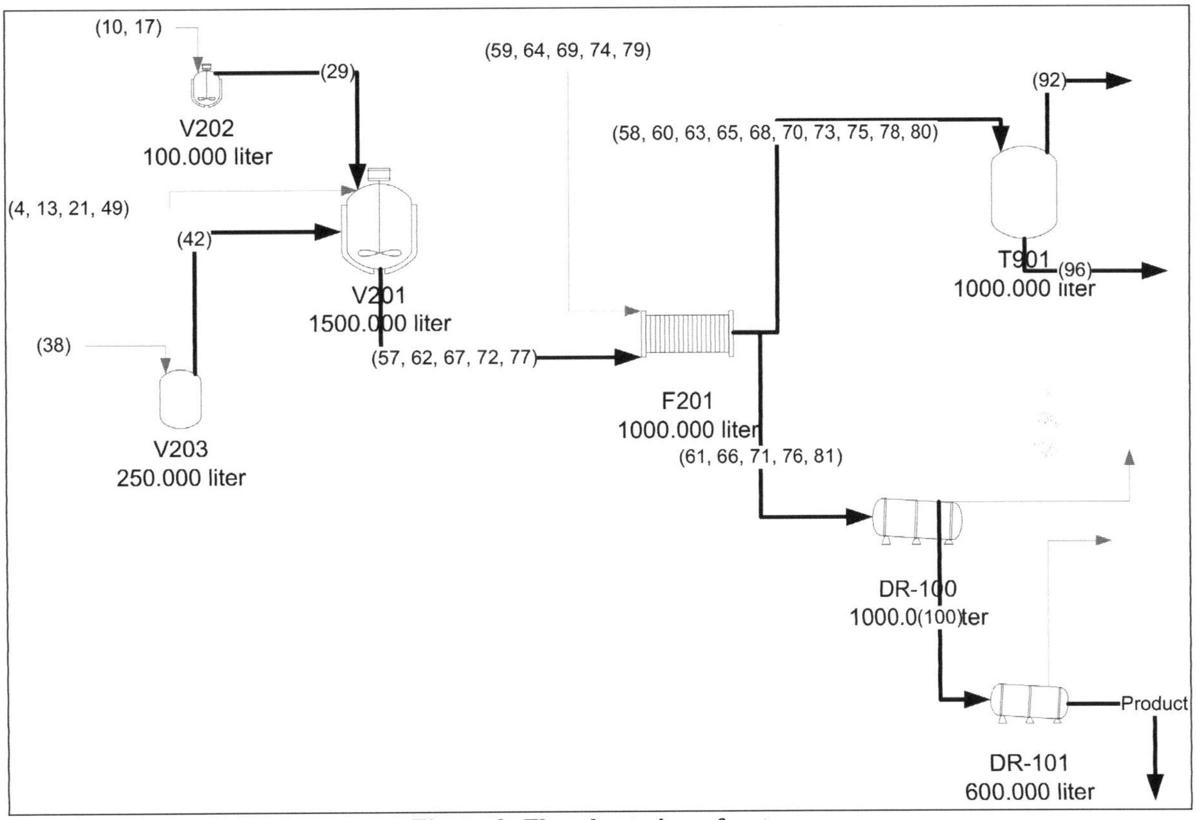

Figure 3: Flowsheet view of a step

Control strategies in the batch industries are at once simpler and more complex. They are simpler in that operations focus on one activity at a time, rather than the entire process. But the complexity arises with the addition of dynamic behavior everywhere. If you need to tweak the last few percentage points out of a fermentation operation, then you need a multivariate, dynamic controller that monitors conditions in the fermenter and keeps the process on the appropriate trajectory while adjusting nutrients, temperature, etc. Multivariate, dynamic control strategies are making their way into many complex batch operations where quality considerations are a must.

Another element is the visualization of batch processes. Depending on the purpose and the level of development, batch processes can be drawn or described many ways:

- Figure 2 - The basic set of transformations (A -> B -> C -> Product)
- Table 1 - The text "recipe" or instructions for conducting these transformations
- A block diagram, describing the sequencing of operations
- **Error! Reference source not found.** - An equipment-centered or plant layout view for process fit and logistics studies
- Figure 4 - A Gantt-chart view of an individual process or a collection of processes running in a given facility to adjust schedule operations

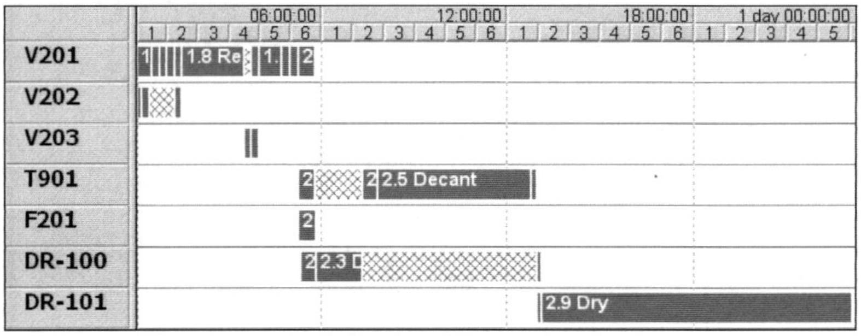

Figure 4: Schedule view (Gantt chart) of a step

There are many other process visualizations, including standard stream reports, materials balances, and cost-of-goods analyses. Note that these all need to represent the time aspect of batch operations in some way. For example, cost-of-goods analyses must account for labor and other operating expenses, based on the time the process consumes.

On the business side, there are a number of interesting aspects of this visualization that offer good educational opportunities—for engineers in particular—in terms of batch process design.

Engineers are not the leaders of many of these businesses. In the pharmaceutical world, it is typically chemists and biochemists who develop the initial process routes and shepherd them through the development cycle, where engineers are more heavily involved. In formulations areas, it is pharmacists or the ink-and-dye experts who design the basic process. As a result, the final process that ends up in manufacturing frequently looks very similar to the process that ran on the lab bench of the original chemist. Another reason why this occurs is that in pharmaceuticals and other regulated industries, the requirement to provide materials for clinical trials locks the business into the basic process quite early in the development cycle. Significant changes to the process require proof that businesses produce the exact same product, and this leads to conservatism in terms of advanced engineering optimization of the process at the later stages. This is a circle-of-influence issue; as awareness about manufacturability rises, the value of engineering input into the Route Selection stage also increases.

Specialty chemical is much closer to familiar territory for chemical engineers, and as a result, many concepts of process optimization have made their way into that arena.

Again in pharmaceuticals and other regulated industries, the level of documentation required is another difference or challenge in batch process design. In fact, most companies call out high-quality documentation as one of the primary deliverables from the development process. This aspect, however, is not taught as part of the standard engineering curriculum - possibly not in any discipline. The FDA and its worldwide counterparts require higher and higher levels of quality and verification throughout the development of new drug products and their processes. It is often quite an education to see the impact of these rules first hand.

Whether the development timeline is measured in months or years, another key element is the required level of collaboration across many disciplines in order to develop a high-quality process, produce trial materials, and create the required documentation. People are involved from many layers of the company as well as every imaginable discipline: chemists, biochemists, lab technicians, chemical & biochemical engineers, pharmacists, formulators, operators, controls engineers, quality control analysts, regulatory filing experts, procurement specialists, contract manufacturers, and many others. These people must all work together and not just on one project, but across the portfolio of projects over time. In addition, these projects will pass across many desks and lab benches. People must understand the impact of their actions and decisions on the rest of the team and project. Those who have the skills and aptitude to work in a highly interactive environment are bound to do well, and they just may have that ability to influence outside their formal area of responsibility.

4 Where do Chemical Engineers play?

Engineers will continue to play important roles in the traditional areas of skill in scale-up, optimization, and operation of chemical and biochemical processes. Beyond these core competencies, chemical engineers have the opportunity to influence batch process development throughout the batch process life cycle.

At the earliest stages of development, engineers need to add their voice to the definition of the best route for new products. Ideally, engineers work hand-in-hand with their colleagues to develop robust processes at the outset of route selection. The most influential engineers may also find their influence moving further upstream, into discovery operations, leading to selection of the best candidate compounds, based on their likely processability, manufacturability, and controllability.

Quality by Design represents an excellent arena for chemical engineers to become involved in pharmaceutical process design. The FDA is asking pharmaceutical companies to provide clear documentation on the “how” and “why” of key process variables. They are looking for proof—whether by experiment, model, or industry practice—that the process by which new drugs are made is under control. From the perspective of chemical engineering, the FDA is asking for a clear definition of the design and operating space for the process. Not only should they know how to run at nominal conditions, but operators should be able to react to process upsets and still produce high-quality product within the design space.

5 Educational perspectives

How does all this relate to the education of new chemical engineers? From a basic perspective, schools should continue teaching conceptual design of chemical processes. They should continue expanding the variety of projects under study in the engineering curriculum, while exploring advanced control strategies and their impact on time-varying operations, such as those found in batch operations.

In light of the larger business in which batch process design resides, engineering programs need to expand their reach into how engineers can positively influence the work of their colleagues from all over the sciences and business. Design projects should be set up to include business students, chemists, and other engineers in order to provide early exposure to the variety of perspectives that other disciplines bring to the same set of questions.

Schools should also teach project management; don’t assume that overwhelmed students will figure it out on their own. My experience is that they either do everything at the last minute (Student’s Syndrome) or they create more work for themselves than necessary (Parkinson’s Law).

In addition, schools should offer courses and seminars in the regulatory environment. It is not only the FDA that heavily influences how engineers do their work, but the EPA and many other bodies play a role in engineering activities. They can be looked upon as impediments to growth—or we can view these regulations as opportunities to solve interesting problems in the context of the larger public need. Encourage students to understand why and how these regulations are created, both the scientific view and the political view.

6 AspenTech support for Batch Process Development

AspenTech has a number of software products that support batch process development activities throughout the development arena. AspenTech provides standard engineering design and simulation tools for traditional process design activities, as well as tools that lay out and describe the process at many levels of detail—from the chemists' route description through the fully detailed set of operating instructions that might be handed to a plant operator or passed into plant automation software. A selection of these applications includes:

- **Aspen Batch Process Developer** (previously Aspen Batch Plus®) for batch process development from the perspective of the recipe or operating instructions. This tool has a wide range of capabilities with respect to the recipe from material balances to emission calculations (EPA requirement) to basic campaign scheduling.
- **Aspen Batch Distillation** (previously Aspen BatchSep™) for design and simulation of batch distillation, batch reaction, or reactive distillation.
- **Aspen Reaction Modeler** (included with Aspen Batch Distillation) for prediction of kinetic parameters, based on experimental data.
- **Aspen Solubility Modeler** (included with Aspen Properties®) for calculation of NRTL-SAC parameters for new chemical entities and conducting virtual solubility experiments (solvent selection; solvent swap; anti-solvent selection).
- **Aspen Chromatography®** for design and modeling of chromatography columns, including adsorption parameter estimation from isotherms.
- **Aspen Process Manual™** is an encyclopedia of process expertise that applies to both batch and continuous operations.
- **Aspen Process Tools™** are a set of specialist calculators and tools for quick predictions in crystallization, drying, and many other operations that are key to detailed batch process design.

All of these tools can be used within the context of teaching batch process development. Aspen Batch Process Developer would be best used in discussions that look at the overall development environment since it is used by our customers in many aspects of the process development workflow. The other products tend to be point solutions that would be valuable in expanding on specific elements of modeling and design courses.

AspenTech offers a large portion of software to the academic community for a low fee. Please see http://www.aspentech.com/corporate/university/index.cfm for more information.

7 Conclusion

While batch processes are a common element in chemical engineering education, the larger business context in which batch process development occurs is a key element for growing the engineering curriculum. It's important for schools to grow students' abilities in project management, collaboration, interdisciplinary understanding, and the impact of regulation.

626e Systematic Methods for Process Synthesis and Design

Lorenz T. Biegler, Department of Chemical Engineering, Carnegie Mellon University, 5000 Forbes Ave., Pittsburgh, PA 15213

Over the past four decades, process design has evolved into a systematic approach to synthesize, develop, improve and evaluate a chemical process - both from technical and economic viewpoints. This approach can be viewed from a number of different levels that include a hierarchy of alternatives, thermodynamic insights and pinch concepts, heuristics that guide decision-making and especially mathematical models for analysis and optimization. Moreover, these strategies have evolved and matured to a level suitable for undergraduate education.

This talk outlines this systematic decision-making framework for design education that combines the above concepts. A number of case study projects are presented along with an outline of the course sequence.

626f Chemical Process Design: An Enduring Philosophy for a Changing Profession

Richard Turton and Joseph A. Shaeiwitz. Chemical Engineering, West Virginia University, 453 Engineering Sc. Building, P.O. Box 6102, Morgantown, WV 26506

The purpose of the Chemical Process Industry (CPI) is, and has always been, to convert lower value raw materials to higher value products. As with all optimization problems, the optimal solution depends on the objective function and the constraints imposed on the problem. A century ago, these constraints were dominated by technology and economics, with few governmental restrictions. Over the years, the impact of the CPI on the environment has become clearly evident. These environmental concerns add additional constraints to the optimization problem, just as improvements in processing technology, materials of construction, catalysts and increasing raw material prices have imposed or relaxed other constraints. In addition, more efficient solution algorithms, faster computers, better simulators, more accurate design tools, and improved design algorithms have expanded the feasible region over which we search for solutions.

A significant change in the type and scale of chemical processes in the US over the last 20 years has led to greater emphasis on batch processing. With these changes, chemical engineers must reconsider some traditional design heuristics, such as those associated with specialty products that differ from those associated with commodity chemicals. These heuristics include packaging, plant layout, product scheduling, unit costing, prototyping (for medical devices and systems), scale-up, and economies of scale. There is no doubt that the number and complexity of issues surrounding the decision to invest in a new process or to optimize an existing process have increased. Nevertheless, the methodology to optimize a chemical process has remained essentially the same. The need to be able to focus on the "big picture" and not get bogged down in all the details is still of paramount importance. Our emphasis at WVU has been to focus on the process as a whole (process design) with less emphasis on equipment design (plant design). It is the presenters' philosophy that no matter how the technical and political landscape changes, the fundamental aspects of process evaluation, process design, and process optimization will remain the same, but the constraints and tools will change.

626g Teaching Design from An Industrial Perspective: The Practical Problems of Practical Problems

Gavin P. Towler, UOP, 25 E Algonquin Rd, Des Plaines, IL 60017-5017

Senior design is usually the course where students get the best insight into the industrial practice of chemical engineering during their undergraduate studies. Most schools make a good effort to ensure that the senior design course provides a realistic experience of working on an open-ended project of practical significance, and many schools rely on adjunct faculty to bring an industrial flavor to the course. Industrial employers have a strong interest in improving the quality of undergraduate education and many are willing to work with universities to enhance the design experience.

There are several problems that instructors commonly encounter when trying to create realistic design projects:

• Students usually have a limited background in solids handling, batch processing, multiphase flow, mixing, and other aspects of unit operations that are very important in industrial design and scale-up.

• Very few schools introduce students to industry codes and standards and show how these are used to simplify the design process and ensure safe, operable designs.

• Students usually have not been taught much about safety, environmental impact and quality control, which have major implications for most designs.

• The problem statement must leave room for the students to demonstrate creativity, while ensuring that they have a reasonable chance of accomplishing something significant in a short amount of time with little or no experience.

• The students probably are not familiar with computer tools for design and costing, even if these are available on campus.

Based on conversations with industrial colleagues who teach design, and my own experience at Northwestern, the paper will give suggestions on how the design course can better prepare students to enter the process industries.

626h Panel Discussion on a Century of Chemical Engineering Design Education

Jeffrey J. Siirola, Eastman Research Division, Eastman Chemical Company, PO Box 1972, 200 South Wilcox Drive, Kingsport, TN 37662-5150, Warren D. Seider, Chemical and Biomolecular Engineering, University of Pennsylvania, Philadelphia, PA 19104, Edward L. Cussler, Chemical Engineering and Materials Science, University of Minnesota, 151 Amundson Hall, 421 Washington Ave. SE, Minneapolis, MN 55455, Jonathan M. Vinson, Aspen Technology, 200 Wheeler Rd, Burlington, MA 01803, Lorenz T. Biegler, Chemical Engineering, Carnegie Mellon University, 5000 Forbes Ave, Pittsburgh, PA 15213, Richard Turton, Chemical Engineering, West Virginia University, 453 Engineering Sc. Building, P.O. Box 6102, Morgantown, WV 26506, and Gavin P. Towler, UOP, 25 E Algonquin Rd, Des Plaines, IL 60017-5017.

Panel Discussion on the papers presented in this session.

Separations: A short history and a cloudy crystal ball.

Phil Wankat
Purdue University

Abstract
Separations have played a major role in the history of chemical engineering and will continue to be important. Since separations have always been the major cost item in the process industries, they have always been a critical key to successful commercialization. Many separation processes such as distillation, adsorption, chromatography and filtration were used long before they were understood. First, the history of separation processes will be reviewed. Next, we will discuss the current teaching of separation processes. The final part of the talk will explore predictions for the future of applications, teaching and research funding in separation processes.

Introduction

Separations have always been and will continue to be critically important in the processing of chemicals. It is common to note that 40 to 70 % of both capital and operating costs in industry are due to separations [1, p. 1]. It has been estimated that 15 % of the world's energy use is required by separations [2]. Because of its industrial importance separations have always played an important part in chemical engineering education and in the chemical engineering literature.

History of Separations

The beginning of separations apparently occurred before recorded history. Egyptians used filtration to filter grape juice over 5000 years ago [3, pp. 89-90]. Aristotle mentioned that pure water can be obtained by evaporating sea water [4, p. 16]. A combination of coagulation of impurities, evaporation, and crystallization used for salt manufacture were commonly in use by the 16th century [3, p. 90; 4, pp. 21; 5, pp. 229-233]. Similar practices were still in use in India in 1980 [5, p. 233]. Pressing, evaporation and crystallization were commonly in use for sugar production by the 16th century [4, p. 23].

Distillation, particularly batch distillation, has a long history. Mesopotamian clay distillation vessels with lids shaped to collect the condensed volatile distillate have been dated to ~3500 BCE [6]. Alchemists in the first century AD in Alexandria used a variety of simple batch stills or retorts [4, p. 16]. The alembic still was invented by Jabir ibn Hayyan (aka Geber) in the late 8th or early 9th century. Similar stills are currently in use in some whiskey distilleries and for distilling rose oil [6]. By the 14th century the production of strong alcoholic drinks had become an industry [4, p. 18]. The first books on distillation were Hieronymus Brunschwig's *Liber de arte distillandi,* in the early 1500's and Walter Ryff's *Das New gross Distillier Buch* in 1545 [7]. Petroleum distillation was started in England in the 17th century and coal tar distillation was first patented in 1746 [4, p. 35]. Fractionation of coal tar into naphta, kerosene, lubricating oil and paraffin was patented in England in 1850 [8]. The first oil refinery constructed in 1860 in Pennsylvania used simple batch stills and collected wide-boiling fractions as the distillation proceeded. Horizontal stills with improved performance were used in the early 19th century for alcohol purification. An improved still was developed in 1818 by Cellier in France who developed a vertical bubble plate still for alcohol purification [4, p.35]. In his *Handbook of Chemical Engineering,* published in 1901 and in an enlarged edition in 1904, George Davis clearly developed the unit operations idea (not by this name) for distillation [9, pp. 100-103]. Before this, the distillation of each chemical was studied separately. It is notable that the 2nd edition of C. S. Robinson's, *The Elements of Fractional Distillation,* [10] has elements of both the unit operation generalization and individual chapters for distillation of a number of chemicals such as ethanol. By the fourth edition, extensively revised by Ed Gilliland [11], the unit operations approach dominated and distillation of specific chemicals were relegated to examples. *The Elements of Fractional Distillation* may have been the first chemical engineering book to also become popular with a non-technical audience – it was very popular with bootleggers in the 1920's and 1930's [12, p. 85]. Batch distillation was important enough to attract the attention of Lord Rayleigh who apparently did the first theoretical analysis of the method [13].

The first continuous distillation appears to have been developed by Aeneas Coffey in 1830 who developed a vertical perforated plate column for alcohol purification [4, pp. 35-36; 13a]. This still was equipped to preheat the feed by exchange with the condensing distillate and the bottoms. In 1900 vertical perforated plate columns quite similar to modern equipment were introduced for distillation of tar [4, p. 36]. Packing was apparently first employed in 1820 and was patented in 1847 [14]. The problem of breaking the ethanol-water azeotrope was solved by Young in 1902 with a batch, azeotropic distillation process using benzene as the entrainer to produce the first observed ternary azeotrope as the distillate product. The batch process was converted to a continuous azeotropic distillation by Keyes in 1928 [15].

In the early 1920's petroleum refineries had not adopted more modern fractionation systems and were using horizontal stills directly heated on the bottom in conjunction with partial condensation to distil petroleum. These systems were not very efficient and considerable redistilling was required. Modernization of distillation in refineries occurred in the 1920's when W. K. Lewis was hired as a consultant by the Standard Oil Company of New Jersey and introduced vertical fractionation systems [8, pp. 305-306]. By the 1920's and 1930's the schematics of continuous distillation columns in textbooks [10; 16; 17] and in Perry's Handbook [18, section 12] look fairly modern except that valve trays and structured packings had not been invented yet. Heat recovery from distillation was common by 1923 [16, p. 575]. The histories of distillation equipment, distillation control, and azeotropic and extractive distillation were reviewed by Fair [14], Buckley [19], and Othmer [15], respectively, for the 75th anniversary of AIChE.

Theoretical analysis of continuous distillation was first achieved by Sorel, who was interested in the distillation of alcohol [20]. Sorel's method is accurate, but confusing and laborious since a trial-and-error calculation was required on every stage. It was analyzed graphically without trial-and-error by Ponchon [21] and Savarit [22] independently. Lewis [23] realized that in many cases the vapor and liquid flow rates are approximately constant – if they are assumed to be constant Sorel's trial-and-error procedure is not required. The simplified Lewis method was converted to a graphical method by McCabe and Thiele [24]. Because McCabe-Thiele plots clarified the physical reasons why column distillation work, the method was rapidly adopted. While no longer used for design, McCabe-Thiele diagrams are commonly used to teach distillation. Solutions for multicomponent distillat ion are much more complicated, particularly if there are non-distributing light and heavy non-keys. Initially, stage-by-stage methods were adapted to multicomponent distillation [25; 26], but closure remained a problem. Numerous computer solution methods were developed after Amundson and Pontinen [27] realized that distillation equations could be conveniently solved after they were put into matrix form. One of the more robust and common methods still used in commercial simulators was Naphtali and Sandholm's [28] linearization of all the equations. The history of distillation models was reviewed by Holland [29] for AIChE's 75th anniversary.

Unlike distillation, which developed gradually over centuries, practical application of absorption appears to have been developed solely by a single person in 1836 [4, p. 29]. William Gossage used an old windmill as a tower to absorb HCl in a downward flowing stream of water. The column was packed with gorse and brushwood. This inspiration soon led to towers packed with various materials such as twigs, broken brick, coke and stone to absorb HCl. A next step was the development of high efficiency CO_2 absorption towers by Ernest Solvay for his Solvay process [4, pp. 29-30]. Theoretical analysis of absorption was facilitated by the development of the two-film theory of mass transfer [30].

Crystallization from solution was one of the key tools of the alchemists [31], and remains partly art. In 1878 Gibbs studied the thermodynamics of growing crystal surfaces at equilibrium and realized that thermodynamics was often not sufficient to explain the crystal growth [31]. McCabe [32] found that the deposition rate/unit area is often linear in supersaturation and deposits grow at a uniform rate. Unfortunately, McCabe's ΔL law often does not hold [31]. Industrial scale crystallizers in 1934 [33] did not look very different than many modern crystallizers. The important theory of crystal size distribution was developed by Randolph and Larson [34]. Hulbert [31] reviewed crystallization for the 75th anniversary of AIChE, and more recent advances in crystal engineering are reviewed by Doherty [35].

Membrane filtration developed at least as early as 1600 BC when the Arawak people of the West Indies used porous stone filters to purify drinking water [36]. With this exception, the development of membrane separations is almost unique since the science was developed before practical applications. The first studies of membrane phenomena were done by Abbé Nollet in 1748 who studied permeation through a semipemeable membrane [37, p. 82]. In 1855 Fick studied diffusion and developed the laws of diffusion still used to study membranes [37, p. 82]. Thomas Graham studied dialysis in 1854 [38] but it was not until 1944 that Kolff and Beck developed a commercial dialyzer – the artificial kidney [37, pp. 95-97]. Graham studied gas separations in 1863 [39] but it was not until 1954 that Kolff and Balzer developed a membrane lung oxygenator that was improved by the work of Clowes and coworkers [37, pp. 131-133]. Pauli developed electrodialysis in 1924 [36] and the multicell electrodialyzer was developed in 1940 [37, pp. 98-99; 40], but electrodialysis did not become practical until the 1950s with the development of synthetic ion-exchange membranes [37, p. 100]. The *seminal* development that led to large scale commercial applications of membranes for pressure driven systems was the Loeb-Sourirajan method of producing asymmetric membranes with a defect-free thin skin [37, pp. 104-105; 41]. This method rapidly led to commercial reverse osmosis systems in the 1960's [37, p. 104; 42]. Loeb-Sourirajan membranes could also be used for ultrafiltration (UF) if the membranes were not annealed. This led to commercial UF systems, but they were severely hampered by concentration polarization. The eventual understanding of concentration polarization led to the development of flow regimes and membrane modules that allowed for practical applications of UF in the mid to late 1960s [37, pp. 117-125]. After Monsanto developed the Prism membrane separator for hydrogen purification in 1979 [37, pp. 129-133; 43], several other commercial gas permeation systems were developed [42]. Pervaporation can be traced to the work of Graham, but the definitive studies were done by Binning and his co-workers in the late 1950s and early 1960s [44]. Pervaporation was first commercialized in the 1980's for breaking the ethanol-water azeotrope [42].

Adsorption, particularly the use of charcoal to purify water, has been known since Biblical times [45, Vol. 1, p. 82], and was used commercially in 1794 for the clarification of raw sugar [46, p. 1075]. Scheele studied the adsorption of gases on charcoal in 1773 [47, p. 548], and the ability of charcoal to remove odors from air was extensively studied by Hunter in the 1860s [46, p. 1087]. Clay was also extensively used with an early use in "fulling" (the removal of grease from wool – hence the name fuller's earth) and processing vegetable oils, and later applications in petroleum processing with percolation processes [46, pp. 1059-1061]. Thermal desorption including burning the adsorbates off of the adsorbent was the common regeneration method if the adsorbent was regenerated. Solvent recovery with activated carbon followed by steam desorption has been commercially practiced since the 1920s with little change in the basic equipment [45, Vol. 1, p. 73]. Pressure swing adsorption (PSA) developed by Skarstrom [48] at Esso in the 1950s and 1960s allowed for much faster cycles and thus higher productivity. PSA was rapidly developed for air drying, hydrogen purification and air separation. Simulated moving beds (SMB) were developed by Broughton and his coworkers at UOP during the same time frame to solve the attrition and mixing problems that occurred in moving beds [49; 50]. This process is similar to the Shanks process (1841) used to simulate counter-current flow in leaching [51, p. 723-724]. Two major commercial applications of the SMB have been purification of p-xylene and separation of fructose and glucose [45, Vol. 2, chapt 6].

Ion exchange can also be traced to biblical times [47, p. 549]. Scientific studies were first done by Thompson in 1850 using naturally occurring clays [52; 53, p. 163]. The first major application of ion exchange, water softening, occurred early in the 20th century [47, p. 549]. The major advance in ion exchange was the development of synthetic polymeric ion exchange resins in England in 1935 [54]. Synthetic polymer resins were used by Frank Spedding and his co-workers for large-scale chromatographic separation of the rare earths in the Manhattan project during and immediately following World War II [54; 55] and are currently used for almost all ion exchange applications including home water softening. Moving bed systems with intermittent or pulsed solids movement have been used for large-scale ion exchange systems, particularly for water treatment, since the 1940s [45, Vol. 2, pp. 68-76]. Applications of ion exchange for biochemical separations followed the demonstration of the power of ion exchange chromatography [56; 53, p. 163].

Liquid chromatography in the form of column elution chromatography was first developed by Tswett in 1903 [57]. He called the method "chromatography" because he observed colored bands moving down the column. Large scale applications of very similar systems were commercialized in the late 1940s for separation of carotene, xanthophyll, and chlorophyll on an activated carbon column using gradient elution and backwash, and in the 1950s the Arosorb process developed by Sun Oil Co. was used to separate aromatics from alkyl hydrocarbons using silica gel [45, Vol 2, p. 1]. Currently, commercial applications of liquid chromatography are common for bioseparations. Liquid-liquid chromatography (LLC) was developed by Martin and Synge [58] and gas-liquid chromatography was developed by James and Martin [59]. Although very successful in analytical applications, neither of these methods has been successful as a large-scale system. However, LLC led to bonded phases that are used commercially in large-scale systems. Scale-up of size exclusion chromatography (SEC) was much more successful, and SEC was used for large-scale separations shortly after it was invented [60].

Teaching Separation Processes

The key questions remain, What to teach? and, How to teach it? [61; 62]. If we had as much time as was needed, we could teach all the separations in a separations oriented ChE curriculum [62]. Because this curriculum is process oriented and uses separations as the unifying theme, it does not fit into current trends in curriculum development [63; 64], but I think industry would like it.

With an overcrowded curriculum separation methods are not going to receive significantly more time; thus, we must choose which separations to include. In my junior level course I have chosen to cover flash distillation, normal and complex continuous distillation (binary, multicomponent, extractive and azeotropic), batch distillation, absorption and stripping, liquid-liquid extraction, and membrane separation. To cover this significant amount of material, extraction is taught at a purely equilibrium level with no design, and membranes are usually limited to gas permeation (but including all flow patterns). I use my own textbook [65], although there are other good textbooks available [e.g., 47; 66; 67]. Obviously, this choice of material leaves out many important separation processes. Some of these are easily included in senior laboratory (e.g., drying and chromatography) or senior design courses, but the students rarely have the same level of understanding of theory. It is also important, if possible, to have dual-level (graduate and undergraduate) electives available on topics such as particulates [68], rate separations [69], bioseparations [70] or advanced distillation [71].

Since the modern practice of chemical engineering uses process simulators and other computer tools extensively, these tools need to be used in the separation course(s). For distillation and the other equilibrium-staged separations, process simulators are used for design and simulation in industry and thus should be used in undergraduate courses [72; 73]. The particular process simulator used is not critical. Since membrane separators are not simulated in AspenPlus, spreadsheets can be used for membrane systems [65; 74]. Spreadsheets [75], MATLAB and Mathematica [76] are useful for solving problems and helping students understand the equilibrium-staged separation methods. *Use* of these tools requires that class time must be set aside for their use – preferably in a computer laboratory. Use of these tools also helps graduates satisfy ABET's criterion 3k, "an ability to use the techniques, skills, and modern engineering tools necessary for engineering practice." [77] Simulators should also be used in dual level electives [69; 71]. The "lecture" portion of the courses should use well known active learning methods [72; 73; 78] in addition to mini-lectures.

Predictions

Where are the industrial use, education and research funding for separations headed? My crystal ball is cloudy, but I will hazard predictions. These predictions assume that the one great "killer" application that makes a host of existing separation processes obsolete will not appear and that funding for academic research on separations will remain tight despite the identification of a number of high priority research areas in separations [79].

First prediction - Distillation: Distillation will remain the major industrial workhorse and a major, although probably slowly declining, part of education in separations. Education in distillation (including absorption and stripping) will increasingly focus on the use of process simulators. Unfortunately, funding for academic research on distillation will remain anemic in the USA.

> *Rationale – Industrial Use:* Distillation will continue to be the industrial workhorse because: 1. It is trusted. 2. It is understood well enough that existing computer models will produce designs that work in about 80 % of cases [79, p. 25]. 3. Except for extractive and azeotropic distillation, mass separating agents are not required and thus do not need to be recovered. 4. A complete binary separation is possible, which means a component can be recovered with high purity and high recovery. 5. In many cases distillation is the most economical separation process. 6. Currently, 90-95% of all separations in the chemical process industry are done by distillation [1, p. 11].
> *Rationale – Education:* Because of the broadening of positions that graduates accept, most schools want to prepare students for jobs outside the traditional chemical and petroleum industries. Thus, there is considerable pressure to teach other separations, but additional time is rarely allocated to separation processes. Process simulators will be used because they prepare students for industrial practice, they are now readily available at most schools, they are supported by textbooks, ABET encourages the use of modern tools, and they help students learn.

Rationale – Research Funding: The USA funding agencies have to a considerable extent apparently decided that distillation is a known art and that companies or Fractionation Research Inc. (FRI) should conduct any research needed. This reasoning ignores that even small advances in distillation can be economically important, and that a paradigm shift, although perhaps unlikely, would have enormous economic impact.

Second prediction: Mechanical separations such as filtration, centrifugation and settling will continue to be ignored in the ChE core at most schools although they will remain critically important in industry. Funding for research in particulates will remain reasonably secure.

Rationale – Industrial Use: Because unwanted solids must be removed and many products are sold as solids, particulate separations will remain industrially important.
Rationale – Education: Unfortunately, at the time that the engineering science revolution changed chemical engineering education, many steps in handling and processing solids were art not science. These unit operations were often dropped from the curriculum since they were not considered to be scientific. Many schools have added these processes back into the curriculum, but in an elective on particulates instead of in the core. Because of time pressures on the core, mechanical separations are unlikely to be added to the core in a meaningful way.
Rational – Research Funding: The mechanical separations have found a home in the general area of funding for particulates. Although not overly generous, this funding is probably secure.

Third prediction: Membrane separation processes will continue to find industrial applications, but at a slower rate than predicted by researchers. Membrane research will continue to benefit from support that is robust compared to that received by other areas of separation. Membrane separations will become an increasingly common part of separation courses in the ChE core.

Rationale – Industrial Use: In applications where they work well (high selectivity and high flux, commercially available, high purity *or* high recovery but not both is required, minimal fouling occurs and the membrane has a long life) membranes are often the least expensive separation method by far. Unfortunately, fouling, relatively short membrane life, and lack of membranes with suitable selectivity often limit use of membranes even when both high purity and high recovery are not required. Researchers in all areas are always overly optimistic about applications of their new inventions – this appears to be particularly true of membrane researchers.
Rationale – Education: Since industry is using membrane separations more, there is a desire to cover this material. Membrane separators are now included in many textbooks, and the level of presentation is accessible to undergraduate chemical engineering students.
Rationale – Research Funding: Funding agencies appear to believe the myth of a killer application for membrane separators without realizing their fundamental limitations. Thus, membrane separators are the only separation systems that are funded at close to a reasonable level. Since myths seem to live forever, this rate of funding will probably continue.

Fourth prediction: Adsorption, ion exchange and chromatographic separation processes will slowly become more important in industry and will continue to receive modest research support, particularly for biological applications. These processes will be taught mainly at the graduate level. Their lack of coverage at the undergraduate level will continue to serve as a barrier to their wider application in industry. Research funding will remain tight although it will be somewhat more available for biological applications.

> *Rationale – Industrial Use:* Adsorption, ion exchange and chromatographic separation processes can often accomplish separations more economically than other methods. This is particularly true in biological applications where distillation is not applicable. However, because most engineers with a BS degree are unfamiliar with these processes, they will be unlikely to consider sorption separations for new applications [79, p. 16].
> *Rationale – Education:* Since the sorption separations are batch processes that require mass transfer calculations, they are inherently more difficult to understand than steady-state, equilibrium processes. Because many undergraduate chemical engineers have considerable difficulty understanding them, these processes will be taught mainly to graduate students.
> *Rationale – Research Funding:* Money will be available for materials applications to make new sorbents, particularly if the research can be tied to nanotechnology. Biological applications have more sources of funding available than non-biological applications such as gas processing.

Fifth prediction: Crystallization will continue to be used in many industries where it is critically important. However, crystallization will remain an orphan without a home in the core of most undergraduate curricula. Crystallization research is currently underfunded and is unlikely to receive large increases.

> *Rationale – Industrial Use:* Since many products are sold in a solid form, the final processing step is often crystallization. In addition, many products such as salts and other nonvolatile materials use very large-scale crystallization. These processes are not going to disappear.
> *Rationale – Education:* Crystallization can be analyzed as an equilibrium staged separation, but the equilibrium is not the VLE that undergraduates and professors are familiar with. A complete analysis that predicts the crystal size distribution requires a mass transfer analysis coupled with population balances. This material is accessible to undergraduates, but because population balances are usually not covered elsewhere in the undergraduate curriculum, considerable time needs to be devoted to the topic. Because of competing pressures to cover other material, most schools will not carve out this time in the undergraduate core despite a call to make crystalline solids one of the core themes in the curriculum [35]. Thus, thorough analysis of crystallization will only be done in elective courses when there is a professor interested in teaching this material. Most ChE graduates have a weak background in crystallization and solids handling in general [79, p. 19], and this unfortunate condition is predicted to continue.
> *Rational – Research Funding:* Much of the funding for crystallization was based on the promise of applications in space. This source appears to have largely dried up and no large-scale replacement sources have materialized.

Sixth prediction: Extraction will continue to be important in industry and to be covered in undergraduate courses, but not enough time and energy will be focused on the unique extraction design issues in education. No prediction will be made on research funding.

Rationale – Industrial Use: Extraction is very useful for cases where distillation does not work although the ratio of distillation to extraction units in industry is approximately 20 to 1 [79, p. 29]. Many of these applications of extraction such as separation of nonvolatile compounds are industrially important.
Rationale – Education: In many ways extraction is the most idiosyncratic equilibrium-staged separation process, although crystallization is a close second. Important content such as third-phase (or rag) formation and design of different types of extractors receives minimal or no coverage. Complete coverage of the methods used industrially would require a separate course. Because of time pressures on the curriculum, this will not happen in the required core. In addition, most current textbooks do not cover and most professors teaching separations are not familiar with these details.

Closure

Since reactors and separators are the core of chemical engineering, these aspects are important in the history, current practice and the future of chemical engineering. The history of separations helps explain how the current practice of chemical engineering separations and of separations in chemical engineering education evolved, and the history provides a useful, but probably limited, crystal ball to predict the future.

References

1. Humphrey, J. L. and G. E. Keller II, *Separation Process Technology*, McGraw-Hill, New York, 1997.
2. Koros, W. J., reported in article by L. Guterman, *The Chronicle of Higher Education,* pp. A1 and A6 (June 20, 2008).
3. Hougen, O. A., "Seven Decades of Chemical Engineering," *Chem. Engr. Prog., 73* (1), 89 (January 1977).
4. Davies, J. T., "Chemical Engineering: How Did it Begin and Develop?" in Furter, W. F. (Ed.), *History of Chemical Engineering,* Washington, D.C., American Chemical Society, *Advances in Chemistry Series*, **190**, 15-43 (1980).
5. Barker, D. H. and C. R. Mitra, "A History of Chemical Technology and Chemical Engineering in India," in Furter, W. F. (Ed.), *History of Chemical Engineering,* Washington, D.C., American Chemical Society, *Advances in Chemistry Series*, **190**, 227-248 (1980).
6. RT, "Distillation through the Ages," *Chemical Heritage, 25* (2), 40 (Summer 2007).
7. Stanwood, C., "Found in the Othmer Library," *Chemical Heritage*, *23,* (3), 34 (Fall 2005).
8. Gornowski, E. J., "The History of Chemical Engineering at Exxon," in Furter, W. F. (Ed.), *History of Chemical Engineering,* Washington, D.C., American Chemical Society, *Advances in Chemistry Series*, **190**, 303-311 (1980).
9. Freshwater, D. C., "George E. Davis, Norman Swindin, and the Empirical Tradition in Chemical Engineering," in Furter, W. F. (Ed.), *History of Chemical Engineering,* Washington, D.C., American Chemical Society, *Advances in Chemistry Series*, **190**, 97-111 (1980).
10. Robinson, C. S., *Elements of Fractional Distillation,* 2nd Ed., McGraw-Hill, New York (1930).
11. Robinson, C. S. and E. R. Gilliland, *Elements of Fractional Distillation,* 4th Ed. , McGraw-Hill, New York (1950).
12. Weber, H.C., "The Improbable Achievement: Chemical Engineering at M. I. T.," in Furter, W. F. (Ed.), *History of Chemical Engineering,* Washington, D.C., American Chemical Society, *Advances in Chemistry Series*, **190**, 77-96 (1980).
13. Rayleigh, Lord (J. Strutt), "On the Distillation of Binary Mixtures," *Phil Mag.*, *4* (23), 521 (1902).

14. Fair, J. R,. "Historical Development of Distillation Equipment," *AIChE Symposium Series, 79* (235), 1 (1983).
15. Othmer, D. F., "Azeotropic and Extractive Distillation," *AIChE Symposium Series, 79* (235), 90 (1983).
16. Walker, W. H., W. K. Lewis and W. H. McAdams, *Principles of Chemical Engineering*, McGraw-Hill, New York (1923).
17. Badger, W. L. and W. L. McCabe, *Elements of Chemical Engineering,* McGraw-Hill, New York (1931).
18. Perry, J. H. (Editor-in-Chief), *Chemical Engineers' Handbook,* McGraw-Hill, New York (1934).
19. Buckley, P. S., "History of Distillation Control," *AIChE Symposium Series, 79* (235), 46 (1983).
20. Sorel, E., *La Retification de l'Alcohol*, Gauthier-Villars, Paris (1893).
21. Ponchon, M., "Graphical Study of Fractional Distillation (in French)," *Tech. Moderne, 13,* 20 and 53, (1922).
22. Savarit, R., "Definition of Distillation, Simple Discontinuous Distillation, Theory and Operation of Distillations Columns (in French)," and "Exhausting and concentrating Columns for Liquid and Gaseous Mixtures and Graphical Methods for their Determination (in French)," *Arts et Metiers, 65,* 142, 178, 241, and 307 (1922).
23. Lewis, W. K., "The Efficiency and Design of Rectifying Columns for Binary Mixtures," *Ind. Engr. Chem., 14,* 492 (1922).
24. McCabe, W. L. and E. W. Thiele, "Graphical Design of Fractionating Columns," *Ind. Engr. Chem., 17,* 960 (1925).
25. Lewis, W.K. and G. L. Matheson, "Studies in Distillation. Design of Rectifying Columns for Natural and Refinery Gasoline," *Ind. Engr. Chem., 24,* 494 (1932).
26. Thiele, E. W. and R. L. Geddes, "Computation of Distillation Apparatus for Hydrocarbon Mixtures," *Ind. Engr. Chem., 25,* 289 (1933).
27. Amundson, N. R. and A. J. Pontinen, "Multicomponent Distillation Calculations on a Large Digital Computer," *Ind. Engr. Chem., 50,* 730 (1958).
28. Naphtali, L. M. and D. P. Sandholm, "Multicomponent Separation Calculations by Linearization," *AIChE Journal, 17* (1), 148 (1971).
29. Holland, C. D., "History of the Development of Distillation Computer Models," *AIChE Symposium Series, 79* (235), 15 (1983).
30. Whitman, W. G., *Chem. Met. Engr., 29* (4), 147 (July 23, 1923).
31. Hulburt, H. M., "Perspectives on Crystallization in Chemical Process Technology," *AIChE Symposium Series, 79* (235), 77 (1983).
32. McCabe, W. L., *Ind. Engr Chem., 21,* 30 & 112 (1929).
33. McCabe, W. L., "Crystallization," in Perry, J. H. (Editor-in-Chief), *Chemical Engineers' Handbook,* McGraw-Hill, New York, 1467-1492 (1934).
34. Randolph, A. D. and M. A. Larson, *AIChE J., 8,* 639 (1962).
35. Doherty, M. F., "Crystal Engineering: From Molecules to Products," *Chem. Engr. Educ., 40* (2), 116 (Spring 2006).
36. Krantz, W. B., "Membrane Science and Technology in the 21st Century," *Chem. Engr. Educ., 38* (2), 94 (Spring 2004).
37. Lonsdale, H. K., "The Growth of Membrane Technology," *J. Membrane Sci., 10* (2+3),81 (1982).
38. Graham, Thomas, *Phil. Trans., Roy. Soc, (London), 144,* 177 (1854).
39. Graham, Thomas, "On the Molecular Mobility of Gases," *Tran. Roy. Soc. (London), 153,*385 (1863).
40. Meyer, K. H. and W. Strauss, *Helv. Chim Acta., 23,* 795 (1940).

41. Loeb, S. and S. Sourirajan, "High flow porous membranes for separating water from saline solutions," U. S. Pat. 3,133,132 (May 12, 1964).
42. Baker, R. W., E. L. Cussler, W. Eykamp, W. J. Koros, R. L. Riley, and H. Strathmann, *Membrane Separation Systems – A Research & Development Needs Assessment,* US Dept. Energy, Contract No. DE-AC01-88ER30133 (April 1990).
43. Henis, J. M. S. and M. K. Tripodi, "A Novel Approach to Gas Separation Using Composite Hollow Fiber Membranes," *Separ. Sci. Technol., 15,* 1059 (1980).
44. Binning, R. C., R. J. Lee, J. F. Kennings and E. C. Martin, *Ind. Engr. Chem, 53,* 45 (1961).
45. Wankat, P.C., *Large Scale Adsorption and Chromatography*, 2 vols. CRC Press, Boca Raton, FL (1986).
46. Mantell, C. L., "Adsorption," in Perry, J. H. (Editor-in-Chief), *Chemical Engineers' Handbook,* McGraw-Hill, New York, section 11 (1934).
47. Seader, J. D. and E. J. Henley, *Separation Process Principles*, 2nd Ed., Wiley, New York (2006).
48. Skarstrom, C. W., "Use of adsorption phenomena in automatic plant-type gas analyzers," *Ann. NY Acad. Sci., 72,* (13), 751 (1959).
49. Broughton, D. B. and Carson, D. B., "The Molex process," *Petrol Refiner,* 38(4), 130 (1959).
50. Broughton, D. B. and C. G. Gerhold, U.S. Patent 2,985,589 (May 23, 1961).
51. Treybal, R. E., *Mass Transfer Operations,* 3rd Ed., McGraw-Hill, New York, (1980).
52. Thompson, H. S., *J. Roy. Agr. Soc. Eng., 11*, 68 (1850).
53. Dechow, F. J., *Separation and Purification Techniques in Biotechnology,* Noyes, Park Ridge, NJ, (1989).
54. Ettre, L. S., "Preparative Liquid Chromatography and the Manhattan Project," *LCGC, 17* (12), 1104 (December 1999).
55. Spedding, F. H., Fulmer, E. L, Butler, T. A., Gladrow, E. M., Gobush, M., Porter, P. E., Powell, J. E., and Wright, J. M., "The separation of rare earths by ion exchange. III. Pilot plant scale separations," *J. Am. Chem. Soc, 69*, 2812 (1947).
56. Moore, S. and W. H. Stein, *J. Biol. Chem. 211*, 893 (1954).
57. Ettre, L. S. and A. Zlatkis (Eds.), *75 Years of Chromatography – A Historical Dialogue,* Elsevier, Amsterdam (1979).
58. Martin, A. J. P. and Synge, R. L. M., "A new form of chromatogram employing two liquid phases," *Biochem. J., 35*, 1358 (1941).
59. James, A. T. and Martin, A. J. P., "Gas liquid partition chromatography: the separation and microestimation of volatile fatty acids. Formic acid to dodecanoic acid," *Biochem. J., 50*, 679 (1952).
60. Janson, J.-C. and Dunnill, P., "Factors affecting scale-up of chromatography," in *Industrial Aspects of Biochemistry,* Vol. 30, Part I, Spencer, B., (Ed.), North-Holland/American Elsevier, Amsterdam and NewYork, 81 (1974).
61. Wankat, P. C., R. P. Hesketh, K. H. Schulz, and C. S. Slater, "Separations. What to Teach Undergraduates," *Chem. Engr. Educ., 28* (1), 12 (Winter 1994).
62. Wankat, P. C., "Teaching Separations: Why, What, When, and How?" *Chem. Engr. Educ., 35* (3), 168 (Summer 2001).
63. Cussler, E. L., D. W. Savage, A. P. J. Middelberg, and M. Kind, "Refocusing Chemical Engineering, *Chem. Engr. Prog., 98* (1), 26S (January 2002).
64. Armstrong, R. C., "A Vision of the Curriculum of the Future," *Chem. Engr. Educ., 40* (2), 104 (Spring 2006).
65. Wankat, P. C., *Separation Process Engineering,* 2nd edition of *Equilibrium-Staged Separations*, Prentice-Hall, Upper Saddle River, NJ (2007).
66. Geankoplis, C. J., *Transport Processes and Separation Process Principles (Includes Unit Operations)*, 4th Ed., Prentice-Hall, Upper Saddle River, NJ (2003).

67. McCabe, W. L., J. C. Smith, and P. Harriott, *Unit Operations of Chemical Engineering,* 7th Ed., McGraw-Hill, New York (2004).
68. Peukert, W. and H.-J. Schmid, "Novel Concepts for Teaching Particle Technology," *Chem. Engr. Educ., 36* (4), 272 (Fall 2002).
69. Wankat, P. C., "Using a Commercial Simulator to Teach Sorption Separations," *Chem. Engr Educ., 40* (3), 165 (Summer 2006).
70. Belter, P. A., E.L. Cussler, and W.-S. Hu, *Bioseparations: Downstream Processing for Biotechnology,* Wiley, New York (1988).
71. Doherty, M. F. and M. F. Malone, *Conceptual Design of Distillation Systems,* McGraw-Hill, New York (2001).
72. Wankat, P. C., "Integrating the Use of Commercial Simulators into Lecture Courses," *J. Engr. Educ., 91,* 19 (2002).
73. Dahm, K. D., "Process Simulation and McCabe-Thiele Modeling: Specific Roles in the Learning Process," *Chem. Engr. Educ., 37* (2) 132 (Spring 2003).
74. Coker, D. T., B. D. Freeman and G. K. Fleming, "Modeling Multicomponent Gas Separations Using Hollow-Fiber Membrane Contactors," *AIChE J., 44,* 1289 (1998).
75. Burns, M. A. and J. C. Sung, "Design of Separation Units Using Spreadsheets," *Chem. Engr. Educ., 30,* (1), 62 (Winter 1996).
76. Binous, H., "Equilibrium-Staged Separations Using MATLAB and Mathematica," *Chem. Engr. Educ., 42,* (2), 69 (Spring 2008).
77. http://abet.org/, ABET (2007), "Criteria for Accrediting Engineering Programs. Effective for Evaluations During the 2008-2009 Accreditation Cycle," accessed June 24, 2008.
78. Felder, R. M., D. R. Woods, J. E. Stice, and A. Rugarcia, "The Future of Engineering Education. Part 2. Teaching Methods that Work," *Chem. Engr. Educ., 34* (1), 26 (Winter 2000).
79. Adler, S., E. Beaver, P. Bryan, J. E. L. Rogers, S. Robinson, and C. Russomanno, *Vision 2020: 1998 Separations Roadmap,* Center for Waste Reduction Technologies of AIChE, New York, 1998.

674b CO_2 Recovery: Myth and Legend

R. Bruce Eldridge, Chemical Engineering, University of Texas at Austin, 1 University Station C0400, CPE 4.408, Austin, TX 78712

Carbon dioxide recovery will be the major challenge for the separations community for the foreseeable future. Various separations technologies are being proposed for large scale recovery. The presentation will address the potential for various approaches to be successfully implemented. A basic case study will be presented with technical, economic, and process feasibility issues discussed.

674c Economical Lower Energy Intensive Large Scale Separation Devices

William J. Koros, Chemical and Biomolecular Engineering, Georgia Institute of Technology, 311 Ferst Drive, N.W., Atlanta, GA 30332-0100

Globally, 37% of total energy is consumed for industrial purposes, while transportation, residential and commercial sectors consume the other 63%. Research into options such as fuel-efficient vehicles, efficient mass transit, green housing and e-commerce has provided options to reduce energy intensity in transportation, residential and commercial sectors. On the other hand, relatively little basic research has been done to significantly reduce energy intensity in the industrial sector.

Lower energy intensive separations can have huge impact in industries such as the petrochemical sector where up to 50% of the energy is consumed to perform separation processes. Reduction in separation energy intensity and CO_2 emissions by more than a full order of magnitude is possible if separations are done by minimizing the use of thermally driven processes.

While understandable in thermodynamic terms, implementation of this vision faces huge challenges. Implementing truly advanced separations technology requires new materials, but it is more than a "materials problem". Indeed, creating high performance membranes and sorbents also includes a "materials processing problem". The intersection of these two general types of problems lies at the core of chemical engineering, so this topic is actually a huge opportunity for our community. A spectrum of length scales must be accommodated in advanced separation devices that are useful for large scale feed streams. Challenges and emerging approaches to deal with these challenges to create the next generation of large scale low energy intensity separation devices such as membranes and sorbents will be discussed. Some realistic paths forward will also be suggested.

SIMULTANEOUS REACTION IN SEPARATION

The other papers in this session carefully outline what is already known about separation science. They show that continued improvements will come in an evolutionary way unless we can shift to an entirely new basis. One such basis is chemical manufacture by the simultaneous reaction and separation. In the past, this strategy has been dismissed because the optimal conditions for reaction and for separation are different. Still, this paper will discuss three examples where such processes seem feasible: the dehydrogenation of ethane, the local production of ammonia, and a wearable chemical reactor to avoid dehydration on airplanes. The examples suggest that the possibility of new paradigms combining reaction and separation have the greatest promise for small systems.

BACKGROUND. The earlier papers in this session give accurate and informed descriptions of the current state of separation processes. In particular, in his paper, Wankat argues that distillation will remain the workhorse of chemical separations. He feels that major improvements are unlikely, even though distillation consumes one million barrels of oil per day in the U.S., with an efficiency commonly believed to be about 11 percent. He asserts that this situation won't change much.

I support this assertion. For differential distillation the height of the tower ℓ is given by

$$l = HTU \text{ g } NTU$$

where HTU is the heighth of the transfer unit and NTU is the number of transfer units. The NTU, which is the measure of the difficulty of the separation, will be small when the operating line is far removed from the equilibrium line, that is, when the reflux ratio is big. But a big reflux ratio means a big heat load in the reboiler, which is a chief cause of process inefficiency. Alternatively, the HTU, which measures the equipment's efficiency, will be small because of structure packing. While this packing will certainly continue to improve a major breakthrough is unlikely.

An alternative way in which separation processes could be significantly improved involves avoiding phase changes, especially phase changes of water, which is the thesis which Prof. Koros advanced. Again, I agree with this goal. Minimizing these phase changes will initially be accomplished mechanically, using presses and screens like those in the paper industry. Further improvements can come from absorption, either by particles or across membranes. Unfortunately, these adsorptive methods may frequently be slower than conventional drying.

<u>Ethane Dehydrogenation</u>. If these strategies fail, we may try to combine reaction and separation. One target example may be the dehydrogenation of ethane, which is run at atmospheric pressure but high temperature, frequently over 800°C. The catalytic reactions which take place under these circumstances are often believed to be limited by equilibrium. Thus the dream is to have some sort of process which removes one product but not the other components. For example, for ethane dehydrogenation, where the reaction is

$$C_2H_6 \text{ É } C_2H_4 + H_2$$

We can imagine a membrane selective for hydrogen, which would continuously remove the hydrogen and hence allow all of the ethane to be converted to ethylene. This type of process has been a dream supported by scattered experiments for at least 20 years. To date, efforts to realize this dream have centered on the flux and the cost of the membrane. Frequently, the target is a cheap, thin palladium membrane in order to achieve a very high flux. Some progress has been made by alloying the palladium with metals like silver.

I remain concerned that membranes of this sort will be difficult to use in practice. I am unsure what seals can be used at 800°C and whether these seals will be able to stand the start up and shut down without the membrane cracking. While the process is intended to operate at steady state, the inevitable inhomogeneities and upsets may give more strain than the membrane can stand. Still, work on this tipic will almost certainly continue.

Small Scale Ammonia Synthesis. A second example is the dispersed manufacture of ammonia, not in a central location but for an agricultural coop or one large farm. At present ammonia is manufactured from natural gas. The gas is burned in air to produce hydrogen, nitrogen, and carbon oxides. After those oxides and water are carefully separated, the near stoichiometric mixture of hydrogen and nitrogen is fed to the Haber process reactor, which operates at about 400°C and 150 atm. As in the case of ethane dehydrogenation, the high temperature, which is chosen to insure fast kinetics, also prevents the reaction from going to completion. Conversion is normally on the order of 20 percent. As a result, the conventional process must have both elaborate apparatus for removing the carbon oxides and sophisticated recycles for recovering the unreacted hydrogen and nitrogen.

An alternative possibility more suitable for making ammonia locally might be to assume that wind power drives the electrolysis of water to produce hydrogen. After nitrogen is produced using commercially an available membrane process, the hydrogen and nitrogen are fed at high temperature and pressure to a batch reactor containing both the Haber process catalyst, a solid absorbent for ammonia. We have made scattered experiments that show that the conversion in this batch reactor can exceed 90 percent avoiding both the elaborate separation equipment for preparing the reaction mixture and the elaborate separations and recycles. We pay for this with the batch reactor, operated periodically. The commercial value of this idea is unknown.

<u>Preventing Dehydration</u>. The final example is a still smaller system. I am partially blind in one eye because I have a blood mutant, common to 15 percent of the population, which makes my blood clot more easily than normal. Thus when I became dehydrated on an airplane, I got blood clots which can cause strokes or, in my case, partial blindness. Thus I have wondered about a wearable absorber which would prevent my dehydration. Such adsorbers would imitate the nose of the kangaroo rat, which allows the rat to go for long periods of time in the dessert without sources water liquid. Such a reversible reactive adsorber seems very possible, easily constructed for less than $100. While I know such a gadget must seem bizarre compared to the larger reactors discussed previously, it does suggest that in the future there is a possibility for a changed paradigm for small reaction units, even if none looks likely for large ones.

674e Separations Pedagogy

Ronald W. Rousseau, School of Chemical & Biomolecular Engineering, Georgia Institute of Technology, 311 Ferst Drive, Atlanta, GA 30332-0100

Place Holder for Ron Rousseau

AUTHOR INDEX

AUTHOR INDEX